P9-DHP-726

THE INTERNET CHALLENGE TO TELEVISION

BRUCE M. OWEN

THE INTERNET CHALLENGE TO TELEVISION

HARVARD UNIVERSITY PRESS
CAMBRIDGE, MASSACHUSETTS
LONDON, ENGLAND
1999

Printed in the United States of America

Library of Congress Cataloging-in-Publication Data

Owen, Bruce M.
The Internet challenge to television / Bruce M. Owen.
p. cm.
Includes bibliographical references and index.
ISBN 0-674-87299-1 (alk. paper)
1. Television broadcasting—United States—Forecasting.
2. Digital television—Economic aspects—United
States—Forecasting. 3. Digital video—Economic aspects—United
States—Forecasting. 4. Telecommunication—United States—Fore-
casting. 5. Mass media—Audiences. 6. World Wide Web (Infor-
mation retrieval system)
I. Title.
HE8700.8.0826 1999
384.55′0973—dc21 98-39236

FOR ALL MY FRIENDS AND COLLEAGUES AT EI

ACKNOWLEDGMENTS

My deepest appreciation goes to the "beta testers" who read drafts of the manuscript and offered comments and suggestions, and to others who patiently explained digital and other mysteries: Michael G. Baumann, Marjory Blumenthal, Robert W. Crandall, Robert Czechowski, Harold Furchtgott-Roth, Peter R. Greenhalgh, Thomas G. Krattenmaker, David L. Nicoll, Roger G. Noll, Peter B. Owen, Gregory L. Rosston, Robert N. Rubinovitz, Philip L. Verveer, George Vrandenberg III, and Kevin Werbach.

I am also grateful to my research assistant, Jason Weintraub, who provided invaluable help, particularly with respect to the figures, the references, and the discussions of Internet demographics. Barbara de Boinville and Elizabeth Gretz as usual contributed greatly to the readability of the manuscript.

CONTENTS

PREFACE

Television, a venerable American institution, faces huge changes as a result of new communication technologies and consumption patterns. This book is about television's future.

The widespread use of television sets in the 1950s transformed the way Americans spent their leisure time, interacted with one another, and perceived the world. And television made fundamental changes in the mechanics of electoral politics and in the constitutional balance of powers, because it enhanced the ability of the presidency to influence public opinion. For nearly half a century, the technology and industrial structure of television remained largely the same. But now change is inevitable, with significant cultural, intellectual, and political implications.

The catchword of the new technologies is "digital," a way of communicating in new forms and processing data much more cheaply. Digital technologies have made possible a steep discount on important services—as if college educations suddenly cost less than popcorn. Unfortunately, most things digital have bewildering acronyms, such as SMATV, DTV, and LMDS. This book attempts to make sense of this alphabet soup. (A glossary can be found on page 341.)

The most visible embodiment of the new digital technology is the

Internet, which consists of millions of interconnected computers. The Internet is both a potential form of television and a potential rival to television. The World Wide Web—the major consumer window on the Internet—permits the sending and receiving of video material, though its current quality, form, and context greatly differ from that of standard TV broadcasts. If the Internet is more effective for their purposes, many businesses, advertisers, and families may abandon broadcast television and use the Internet instead.

This book focuses on the important nexus between television and the Internet. Both Internet and TV signals are or can be carried on some of the same communication pipelines, such as fiber-optic cables and satellites. Just as personal computer screens can be used to view ordinary TV broadcasts, television sets can today provide access to the Internet. The distinction between television and the Internet, or between the TV set and the telephone, may disappear. This hypothesis is called convergence. Whether convergence is inevitable, or even likely, is still unclear. At the moment, because television and the Internet are quite distinct, one can readily discuss one without mentioning the other.

The salient characteristic of the Internet for our purposes is that it has limited capacity, or "bandwidth." Television, in contrast, is a great gobbler of bandwidth. The metaphor of the camel that is too large to fit through the eye of a needle suggests the difficulty of fitting television signals onto the Internet. In fact, as things stand, it is impossible to do so. Whether the future will bring smaller camels, larger needles, or better means to stretch things remains to be seen.

Behind convergence lie technological change and large ongoing productivity gains. Both computers and communications today are digital, because the unit cost of each is low and falling rapidly. The cheapest way to get data from here to there involves a combination of computing (digital signal processing) and transmission. The two can be traded off. In recent years computer costs have been dropping even faster than transmission costs. As a result more and more processing goes on at each end of each transmission pipe. Any technology of communication, such as conventional broadcast TV, cable TV, or satellite TV, is some combination of such transmission paths and processing or coding equipment at the ends of each path. The choice of technology determines the form in which information is sent and received.

And it is the attractiveness of that form to consumers and its effectiveness for advertisers that help determine whether society will be willing to pay the costs of adopting the technology.

Another focus of the book is the effect of government regulation. Since 1927 the federal government has dictated the development of much of the communication industry with a heavy hand and, from the point of view of consumers, has done an extraordinarily bad job of it. Many communication services, new or imminent today, could have already been available for years, but for Congress and the Federal Communications Commission (FCC). Politicians and industry incumbents both benefit from regulation, and therefore it is likely to continue and to affect the rate and direction of technological change.

Because of regulation, certain industries and firms have been identified with, or even tied to, particular technologies. For example, television broadcast networks such as CBS or ABC have not been permitted to own cable television systems. When any such technology is threatened with obsolescence, its owners or users naturally seek help from the government that created and constrained them. The instinct of government in these circumstances is to delay the new entrants until the incumbents have a decent interval—perhaps a decade or two—to wind up their affairs.

While communication technology and economics are universal, regulatory and political institutions are not. Even though all the world loves American films and television programs, and even though the Internet is so ubiquitous that it poses threats to national sovereignty, there are considerable differences in the structure of electronic media in different countries. This book focuses on the American context.

In considering the future of television it is useful to understand the physical performance characteristics and costs of various communication technologies. Unfortunately (at least for analysts) these costs are constantly changing. Today's estimates (however accurate) are unlikely to be valid a few years hence. For this reason, I have generally not supplied citations for such numbers, and they should be regarded as merely illustrative.

Finally, this book endeavors to keep the reader clear-eyed. Many of the experimental new communication technologies and services are wonderful, and it seems that it would be terrific if they came to pass. But we know that most will not. The issue is not how terrific they are, but how much they cost and whether there is enough demand to make it profitable for someone to accept the risk of providing them. When considering the broad social, cultural, and political implications of the new digital technologies, it is wise to keep in mind the view of the historian George Daniels (1970, pp. 3, 6): "No single invention—and no group of them taken together in isolation from nontechnological elements—ever changed the direction in which a society was going. . . . [T]he direction in which the society is going determines the nature of its technological innovations. Habits seem to grow out of other habits far more directly than they do out of gadgets."

Television itself caused fundamental change in American life, and seems to be an exception to Daniels's proposition. But few innovations can make such a claim. It is clear that digital technology has had and will have important effects on society and on television. Further changes are inevitable. But despite the hype, it is by no means clear that the Internet, as distinct from digital technology, will revolutionize television, much less the American household.

The book is organized into an introduction, four main parts, and a conclusion. Part I (Chapters 1–2) is intended to convey basic concepts and fundamental ideas. These chapters are somewhat condensed and contain a preview of the remaining portions of the book. Part II (Chapters 3–8) covers the history and regulation of analog media such as radio and television broadcasting. Part III (Chapters 9–12) aims to explain the economic tradeoffs involved in designing digital media such as the Internet. In many ways this part is the core of the book. Part IV (Chapters 13–17) surveys the most promising of the new digital media and seeks to explain the economic tradeoffs and business strategies that will help determine their success. Part IV should not be interpreted as an attempt to "pick winners" in the marketplace. Nevertheless, readers interested in how businesses react to the opportunities and challenges of new technology may find the stories in these chapters useful even after circumstances have changed.

Washington, D.C.
June 1998

PART ONE

THE BASICS

CHAPTER ONE

TELEVISION, THE INTERNET, AND CONSUMER DEMAND

After a half-century of glacial creep, television technology has begun to change at the same dizzying pace as the wares of Silicon Valley. Although computers form and inform important aspects of our popular culture, and increasingly so, television—old-fashioned broadcast television—has been central to pop culture for many years. To understand where television is going, and why, requires a grasp of the economic history of the industry and the effects of technology and government regulation in shaping its organization. Recent developments associated with the growth of the Internet must also be explored. With this history to stand on, the reader can peer into the future and form views regarding the likely effects of television and the Internet on each other and on the future of video media. We can at least make an educated guess, for example, about whether the computer, the TV set, and the telephone are converging, and if so what form that convergence will take. We can begin to fathom what will happen to television as we know it today.

Nonspecialists struggle to follow the new moves in the digital world as the communication titans jockey to survive amid what Joseph Schumpeter called the "gales of creative destruction." It is not merely a game of innovation and investment or obsolescence and decline; it is also a game that is played out in Washington and in media hyperbole. Many are tempted to follow the technology, thinking science will determine the "outcome." Some follow the business leaders, thinking their powerful corporations call the shots. Still others watch the federal government, which makes the rules. But knowing who the players are does not explain the game. The game is an economic one, involving interactions among all the players, including consumers and advertisers, each of whom pursues a disparate goal. Those interactions are governed by certain economic principles, such as the pursuit of self-interest, that can serve as powerful predictive tools. At the same time, the game is far more complex than the sum of the players and their motivations, both because of interactions among the actors and because of the interference of government.

We Spend So Much Time at It, It Must Be Important

Since the 1950s Americans of all descriptions have spent a substantial portion of their lives watching television (see Figure 1.1). Broadcast and cable TV viewing by the average person peaked at 20.4 hours per week (one of several contentious estimates) around 1984, although it has since declined slightly, partly because of video rentals (see Figure 1.2). Television viewing is an activity of all social classes and income levels, but is especially common among the elderly, the young, and the less educated. More households have televisions than telephones. TV watching consumes about 40 percent of leisure time.

Finally, More Options—If You Pay

In the early 1950s, the federal government arranged matters so that nearly every American had three choices of what to watch on television, whether for local or national programs. Those choices became the major TV networks: CBS, NBC, and ABC. By contrast, Ameri-

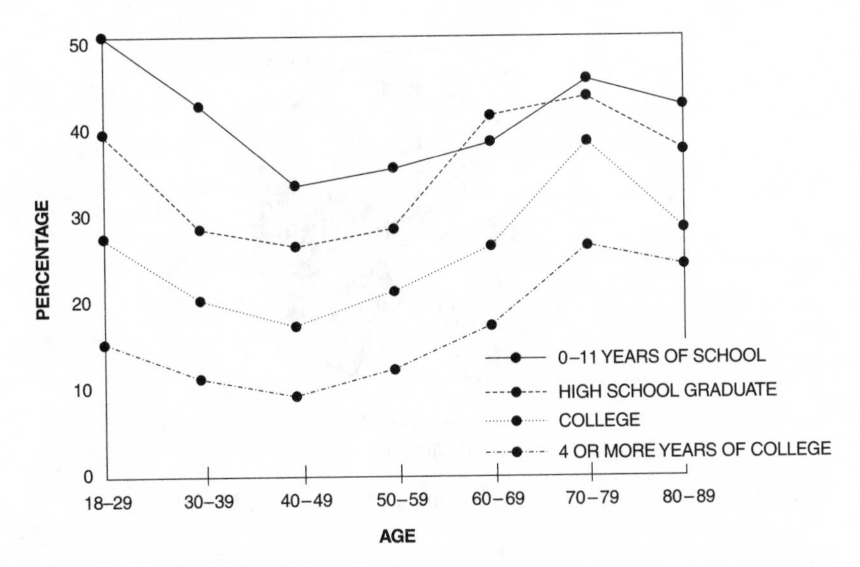

Figure 1.1 Percentage of Americans viewing four or more hours of television per day, by age and education (National Opinion Research Center, 1977–1994)

cans today have access to an average of 12.4 over-the-air local television broadcasts, 4.5 national broadcast networks, a local cable system with an average of 52 channels, four competing satellite broadcasters with up to 200 channels, video rental outlets and catalogs with thousands of titles, and untold dozens of sources of "video-like" content on the Internet. Figure 1.3 details the astonishing recent growth of cable and broadcast networks.

The Telecommunications Act of 1996 sought to end the legal barriers that have kept telephone companies from offering video services and cable companies from offering telephone service. The Federal Communications Commission's increased willingness to permit competition and to make use of market forces in allocating the airwaves will hasten the pace of change. But the FCC will continue to play a crucial role, and that role has traditionally been influenced not merely by the available technological choices but by the very industry it regulates.

From an economic point of view, perhaps the most important development has been the advent of technologies and business plans that permit consumers to pay for programs, voting with their dollars

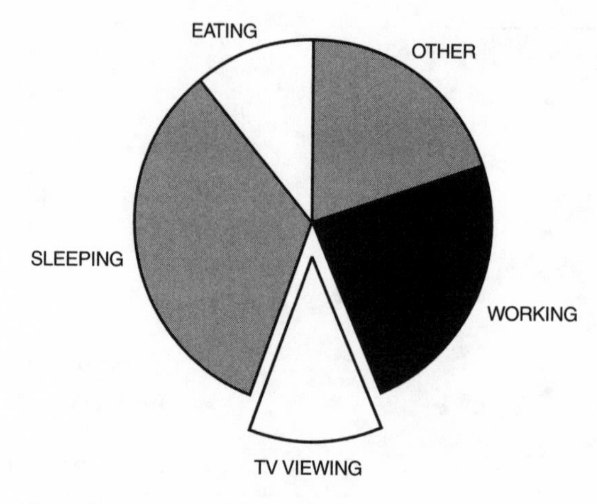

Figure 1.2 Allocation of time: U.S. adults, 1990 (Larson, 1992)

rather than merely with their eyeballs. Until this happened the extent of the television industry could never be greater than the supply of advertising dollars. Consumers, however, tend to place a much higher economic value on programs than do advertisers. One consequence is that in cable television the contribution of subscription fees to revenues far exceeds the contributions of advertising. Allowing consumers to pay vastly increases the quantity of television programming that can be supported, which is why there are dozens of cable networks, even though the audience for most of them is very small.

More than One Kind of Television

There are four types of present and future television.

- *Conventional TV* is dominated by ABC, CBS, and NBC. Together these three networks still capture nearly half of the national prime-time audience.
- *Cable TV* has many more channels and networks and a great deal of specialized programming, all because it can tap consumer dollars. Otherwise, it is similar to conventional TV.
- With *interactive TV,* the viewer decides in advance what programs (usually movies) she will watch, and at what time. Interactive TV includes videotapes rented from a store as well as "video

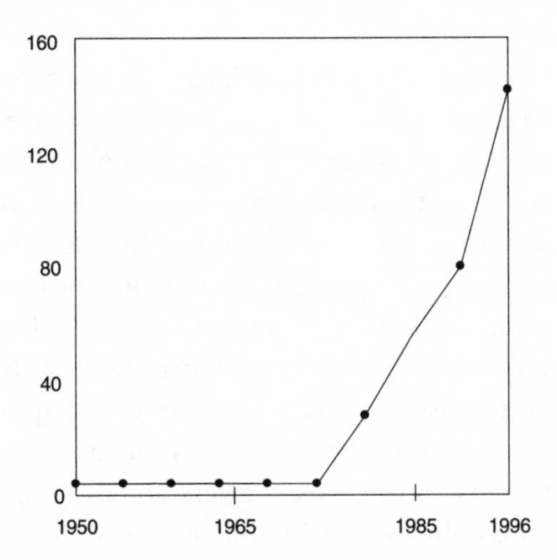

Figure 1.3 Number of broadcast and cable TV networks, 1950–1996 (National Cable Television Association, 1997)

on demand (VOD)" and "near video on demand (NVOD)" delivered by cable, by satellite, or over the air.

- *Internet TV* refers to video images that appear on a computer monitor in connection with World Wide Web pages. These images are just as responsive to the user's demands and commands as hypertext or other Web user interface features. ("Internet TV" may also include two-way switched or routed video telephone services, but that is not our focus.)

Each step from conventional television to Internet television requires, very roughly, an order of magnitude increase in transmission capacity, keeping other things equal.

Conventional television was the television of the 1950s, 1960s, and 1970s. By 1980 most new TV sets were sold with remote controls, which made television "interactive" in the most primitive sense. By the end of the 1980s cable television was available to nearly every American household, and more than half subscribed. Doing so permitted the average subscriber to increase the number of viewing choices by a factor of three to five. Having more choices and a remote control makes it possible to characterize cable television as "more" interactive than conventional television, but the viewing experience remains relatively passive.

Fully interactive television means that the viewer decides not only what to watch but when to watch it. A rented videotape is interactive in this sense. A rented videotape is even more interactive with the user than is VOD or NVOD, because the tape can be stopped, rewound, and fast-forwarded. A step beyond that degree of interactivity requires an increase in the size of the potential library of video products and an increase in the speed of response to user commands; such television does not exist today, unless one counts video games. The closest analogy to fully interactive television is the World Wide Web.

Conventional television is a very different consumption experience from Internet television. For many viewers the first is largely passive. Even with a remote control, most viewers don't change channels unless provoked. It is quite easy to do something else, even several things, while watching such television—in some ways it can be treated as the visual equivalent of Muzak. Internet television requires that one pay close attention. There is nothing passive about it. Like a video game, highly interactive programs on Internet TV require as much focused attention as work or active sports.

There are two reasons for these distinctions among types of television. First, they use different transmission technologies, and they have different capacities and costs. Second, they are, from the consumer's point of view, different products. Although each of these four kinds of television may be regarded as substitutes by consumers, they are not necessarily the *closest* substitutes for one another. Other leisure activities, such as eating out or reading, may be better substitutes for conventional television than is Internet television, for example.

Here Comes the Internet

Equally portentous is the Internet. The Internet can deliver video, albeit at this point poor-quality video. (Bandwidth, understood broadly to encompass the data transmission capacity of a medium, is a major limitation on the Internet as well as on TV broadcasting.) The Internet permits payments by both advertising and viewers, although neither form of payment is yet institutionalized. But more important, the Internet is the first potential substitute for broadcast television as an in-home entertainment delivery medium since the rental videocassette. Those households that are experienced with the Internet may spend as much or more time *with* the computer screen as *before* the television screen. Those who use Internet services spend an average of twenty hours per week doing so, according to one survey, and report that they view less television than they once did.

Television sets themselves can now be wired to devices such as WebTV that function as simple Internet terminals. The TV screen becomes a low-quality computer monitor. While in such use, television sets do not display conventional television signals, although WebTV permits TV programs to appear in a small window on the Web page. The Internet has joined "Doom" and the other Atari/Sega/3DO games as rivals of the broadcasters for TV time. One reason for the success of DSS—the new General Motors/Hughes broadcast satellite television service—is its over two hundred choices. But the choices the Internet can offer are unlimited, or no more limited than those in videotape rental stores and catalogs. Video is not today an important source of demand for Internet access. The important sources of demand are such services as e-mail, chat rooms, and Web browsers. About 40 percent of United States homes are equipped with personal computers, a number that appears to have stopped increasing rapidly. Average hours of weekly Internet use among those who are connected, business and residential, has increased from about three to about seven over the period 1995–1997. The number of e-mail users and the volume of e-mail per user have increased rapidly and are expected to skyrocket in the coming decade. E-mail undoubtedly is the most popular use of the Internet, and for many it is their only use of the Internet.

On the Internet, viewers can tailor their leisure consumption of information and entertainment and advertisers can target their audiences with the same precision as print media. The Internet user can choose among literally millions of sources of information and amusement on the World Wide Web. These choices often are, or could be, rather like television shows—colorful and animated, using images as much or more than words to convey meaning. Like the TV viewer, the Web user is glued to a screen, but the experience usually is more absorbing, because the Web is interactive. Although users Web surf just like TV viewers channel surf, the surfing on the Web is less passive. Although the rate at which data reach the computer screen today is often agonizingly slow, there is some compensation: the user has specifically requested what is on the screen. Today the Web competes with television for some viewers' time, but probably soon it will compete for viewers' dollars as well. That competition may not end happily for conventional television.

Sometimes It's Nice to Be Passive

What is the difference between surfing 200 conventional video channels and 2 million Web pages? The difference is control. The Web browser can control the pace at which content is received and can select the moment at which the scene shifts. Even pay-per-view television, or channel surfing with a remote control and no fellow viewers to accommodate, cannot approach this degree of control. Because of this difference, the Internet content provider can either charge more or offer lower-quality (less costly) content in competing with conventional video services. At the same time, the absence of interactivity is also valued. Much TV viewing is very comfortably passive. Consumer surveys show *decreased* satisfaction with television over the years in spite of great increases in the number of choices and greatly increased ease of use. More choices and remote controls require an often unwelcome investment by the viewer in making and executing decisions.

It is ironic that the Internet is not distinguished from television as a source of information "overload" or "data smog," as David Shenk (1998) calls it. Except for "push," the hot technology of 1997, now apparently moribund, it is television and not the Web that drenches the user in data. This is so in the literal technical sense that television bombards the viewer with at least an order of magnitude more bits of data per second than the computer monitor of a Web surfer. More meaningfully, the Web surfer controls both the content and the pace of arriving information (except when it is *slowed down* by traffic jams on the Internet). The television viewer gets to choose the channel and nothing else. Compared with television, the Internet gives people greater power to control information, reducing the danger of overload.

Studies of leisure time use shed light on demand issues. One by John Robinson and Geoffrey Godbey (1997) provides these findings:

- Since 1965, Americans on average have gained about one additional hour per day of free time.

- Having a television, thirty to forty years ago, was associated with a doubling of the time spent on media consumption as a primary activity. In other words, television was successful because it drew time not only from radio listening and reading but because it drew time from virtually every other daily activity, including sleep.
- Television as a primary activity now occupies about fifteen hours per week of the leisure time of the average American adult. Despite the proliferation of program choices, this percentage has changed little since 1985.
- Television is inexpensive, undemanding, and convenient. It is suited to viewing in disconnected and fretful bouts and with frequent interruptions. It may be habit-forming.
- Despite its vast popularity, TV is not popular:
    - When asked what activity they would preempt if something unexpected came up that required an hour or two of time on a given day, most people say television.
    - Among other activities, TV rates below average in enjoyment: above reading the newspaper, but below grocery shopping and house cleaning.
    - Television, more than any other leisure activity, declined in enjoyment from 1975 to 1995. Newspaper reading dropped by a similar magnitude, but reported viewing did not decrease.
- Users of computers and online services tend to be heavy consumers of print media, and their television viewing is not much different than that of otherwise comparable nonusers.

Robinson and Godbey speculate:

> Interactive entertainment on the Web or elsewhere can never substitute fully for conventional television because it cannot satisfy the need to be entertained passively. We often are not as effective at amusing ourselves, even with all the tools of our community and culture, as we are being entertained by another. . . . Part of the allure of television is *freedom from choice.* It is a respite from an active world. Interactive television . . . may actually be less appealing to people . . . if they must invest more energy and imagination. (p. 312)

Conventional television seems to satisfy a demand for which interactive television is not a very good substitute.

Internet Demand

Less than half of American households own a computer, and at this writing only 15 percent have access to the Internet. (Internet here excludes the proprietary online services, such as AOL, which are used increasingly as Internet access providers. Of course, many persons have Internet access through businesses and schools.) Of those households with access to the Internet, only about 15 percent make heavy use of it. So, allowing for some inaccuracy in the data, it appears that less than 10 percent of American households use or know how to use the World Wide Web as a regular domestic service. Even at the office, Internet usage is chiefly to send and receive e-mail. These are startling data given the enormous attention paid to the Internet by the media and the ubiquity of Internet site addresses (URLs) on commercial materials.

The major reasons for such low penetration, despite all the hype, are the cost and difficulty of setting up and using the equipment. Most people don't own or know how to use a computer, much less want to watch television on it. If a modern PC or even a Macintosh computer were built into new TV sets, most people would not want to buy them. That computing costs will fall is also a given. That the unit cost of digital TV and the Internet will fall also is a given. But it certainly is not a given that someone will design an interface that permits ordinary people—people who are unwilling to learn to program their VCRs—to make effective use of a computer, or that someone will think of an entertainment format that will make ordinary people want to use a computer for that purpose.

Aficionados of personal computers and other consumer electronics find it hard to grasp that many people do not share their intuitive grasp of the digital medium, just as many are flummoxed by the machinery under the hood of a car. According to one report, fully one-third of the customer support calls that Toshiba (a major manufacturer of television sets and other consumer electronic equipment) receives are resolved by having the customer plug the device into an electrical outlet.

Uncertainty over the fate of television cannot be resolved until the creative community discovers the form or forms that make sense of the new digital media; yet the medium itself continues to evolve so rapidly that there is little time to explore its potential. The resulting period of chaos and uncertainty will greatly increase the risks facing investors while providing nearly unlimited opportunities for entrepreneurship and creativity. The situation resembles that of radio broadcasting in the 1920s.

If consumers are willing to pay enough, almost anything may happen; if they aren't, breakthroughs on the supply side will be fruitless. It will be the task of the creative community (perhaps but not inevitably the one that now produces TV programs) to invent formats that take advantage of the strengths of the new medium. One sees this most clearly in the burgeoning and diverse industry of Web page designers. But unless greater strides are made in simplifying interface designs, Internet use will not become a mass medium, no matter how cheap it becomes.

What is the "right" format? Which images of information and entertainment will succeed in making the Internet a video broadcast medium, or a medium that displaces video broadcasts? Underlying content will probably remain much the same: news, weather, pop entertainment, classics, documentaries. The kinds of stories that we tell one another will not change, but the form in which they appear must change. Suppliers of digital content will discover what forms best suit their medium. That is, they will discover what unique combinations of animation, resolution, download time, and other characteristics make the most attractive form of delivery for Internet information and entertainment. There is no reason to suppose that this format is present-day video.

Digital Advertising

Advertisers' aggregate demand for audiences is small compared with consumers' willingness to pay for the traditional video content that attracts audiences. None of the new digital media, whether "conventional" in format, like geosynchronous communication satellites, or "new," like the World Wide Web, will support delivery of mass audiences. Because of their great capacity to offer specialized content, digital media fragment audiences; they do not aggregate them.

Thus it will be easier with digital media to advertise golf balls and harder to advertise soap. Because advertisers pay more for audiences made up chiefly of likely purchasers of their products, however, Web advertising may have a bright future despite fragmentation. The advance of Web advertising will continue to leave room for conventional mass media capable of delivering large audiences to advertisers seeking them.

Conclusion

Television is consumed in vast quantities. Consumer use of the Internet and online services, while growing, remains small compared with TV use. An important factor in consumer demand for television is a desire for passive entertainment. Internet content currently satisfies an entirely different kind of demand. If the Internet is to compete with television as a mass entertainment medium, the industry must invent an interface and format accessible by mass audiences. Even then, no one knows whether there is sufficient consumer or advertiser demand for broadband digital services to support the required massive infrastructure investments.

CHAPTER TWO

SUCCESSFUL MEDIA TECHNOLOGIES

What determines the commercial success of communication media such as television, radio, newspapers, and the Internet? Satisfaction of consumer and advertiser needs is an important condition, but technology matters as well. Technology, however, keeps changing.

Nothing Seems to Last

Video and telephone communications over the past half-century have seesawed back and forth between technologies. At the time of World War II, long-distance telephone calls traveled on copper wires, from one operator to another, in small steps across the continent. After the war, long-distance service was supplied by microwave radio repeater towers all over the country. These microwave radios carried not only telephone calls but also video signals headed for local broadcast stations. Microwave, in turn, was replaced, especially for video transmission, by geosynchronous communication satellites. Satellite technology was then successfully challenged by fiber-optic cables. In spite of these trends, few media die. But each must constantly shift to find its current niche.

Today wireless transmission in various forms again challenges wire for telephone and video service. The seesawing between wires and radios is partly the random walk of technology. Regulatory constraints and competition-induced changes in the emerging technologies also contribute to this back-and-forth phenomenon.

Communication firms often identify with particular technologies rather than with particular products or services. Market divisions are created and maintained by government rules, to be sure, but in many cases the most efficient way to organize production is around a technology rather than the product or service being produced. (Cable companies, for example, tend not to own wireless cable licenses.) One effect of this is to make firms whose technologies are threatened particularly tenacious in resisting change.

One Wire?

In some ways the Internet is a telephone service: the local lines that connect users with Internet service provider (ISP) gateways are telephone wires. Cable systems could also provide this service, but most do not yet. Even when they do, telephone wires may be required. Satellites could do so, but to date there is only one such service, and it too requires telephone company wires. Therefore, if the Internet conveys video services or services that compete with video, the telephone companies may well find themselves in the video business as well.

The prophecy of convergence is this: television sets, telephones, and computers (and the networks that bind them) are or will become the same. The Internet will be all. Convergence implies a fundamental competitive struggle for survival among media, including those that do not compete today. Convergence can also mean something less fundamental and for present purposes less interesting—the tendency of some Web sites (for example, America Online) to use television metaphors and the corresponding tendency of some video sources to offer Web-like content (for example, Bloomberg financial news).

There are literally hundreds of feasible variations of communication technology already available. Each configuration of physical attributes is associated with a technically defined set of creative possibilities for the structure of content. Each has its own set of costs. Together, characteristics and costs determine which media are efficient. Of all the possibilities, only a few media actually come into commercial use, either because they triumph by satisfying consumer demand in the marketplace or because they are defined and protected by government regulation, or simply because their proprietors are lucky.

Will television broadcasting, obviously a successful medium in the analog world, continue to succeed in the digital world, and if so how? If not, will it be digital "multimedia" that drives out television, or something else? (The term "multimedia" is not very precise. What most people seem to mean is a computer that can show some combination of text, pictures, TV images, and sound effects.)

Many perspectives can help us understand successful media.

- How and why did our present media (print, radio, television) overtake rival technologies?
- What are the scientific and technical functions of communication media, and how are these affected by invention and innovation?
- What are the economic factors and processes that determine whether one communication medium or another will be widely adopted?
- How is the act of consuming communicated content related to media success, and how is content itself shaped by and determinative of media forms?
- How does one medium successfully harness the power of government to protect and promote itself at the expense of rivals?

As these questions suggest, no single factor "explains" how one medium succeeds while others wither away. An understanding of successful media must come from several directions. Technology provides us with a set of feasible media forms, and economic analysis can assess their costs. Further, economic outcomes feed back on R&D incentives. That is the supply side. On the demand side there are three forces. First, in the television industry, government is paramount. Through regulation and legislation, and frequently by the invitation of those it regulates, the government defines and constrains what the television market is permitted to provide. (Of course, the government also plays a supply-side role, limiting entry and restricting access to the spectrum.) Second, advertisers and merchandisers have well-understood demands for audiences of various types and sizes. Finally, there is consumer demand—the willingness of consumers to pay for new media services and forms—about which we can say little except through analogies with past behavior.

There are many futures for broadcast television and the Internet. One is convergence. Another is that television and the Internet, though each digital, will keep on their separate ways, joined chiefly by competition for viewers' time and attention and for advertisers' dollars, and not by a melding of transmission paths or consumer equipment. The cost of interactive standard-quality video delivery may be too great to be supported by the demand for the services it can provide to subscribers and advertisers. In that case, even though television and the Internet may share some of the same transmission pipelines, and in certain cases the same terminal equipment, television will continue to be noninteractive. The Internet, lacking content valuable enough to support a broader highway, may continue to struggle along with narrowband transmission into and out of the home. Conventional means of broadcasting (including satellites) may continue as the most cost-effective delivery system for noninteractive programming. If this advantage persists, satellites or wireless cable could gradually displace conventional broadcasting and cable, killing the Internet's chance to be a video mass medium. There are too many possibilities, and too great a rate of change in technology and economics, to make predictions.

Broadcast Communications

Narrowcast and broadcast are the two basic kinds of communication. Narrowcast communications involve one-to-one or one-to-a-few two-way communicants. Broadcast messages are one way, one to many.

Broadcast communications have these special features. First, the content is a public good; that is, production costs do not increase when viewers are added. Therefore, larger audiences mean lower unit costs. Second, the message is valued by consumers in part because it is shared, and near-simultaneity of receipt is an enhancement or feature of sharing.

Broadcast communications can take various forms, each dictated by the technology of the medium, and each influencing the content of the message. A newspaper may broadcast much the same message as television, yet the two forms are very different and make for a different consumption experience. A fundamental distinction among broadcast media is whether the medium is itself a public good. A newspaper, even though it is a broadcast medium, is not in its physical form a public good. Each reader gets his or her own copy. Only the content is shared, not the paper. Similarly, video "broadcasts" on today's Internet are not public goods because identical, separate, streams of packets must be sent to every viewer. In contrast, over-the-air broadcasts, satellite broadcasts, and certain computer network topologies permit public-good transmission of content.

In order to be successful, a transmission medium must perform a given function more cheaply than alternative media or (because of its form) offer an opportunity for the transmission of more valuable content. A brief review of media history illustrates this point.

How We Got Here

In the nineteenth century newspaper publishers discovered that low prices for readers and mass-appeal content could generate advertising revenue that would support all or nearly all of the costs of publication. Magazine publishers discovered the rewards of tailoring content to the tastes of special-interest readers that advertisers prized.

Print media, however, are very inefficient in their reliance on making and transporting a separate, identical copy to every reader. Despite this, and in contrast to early electronic media, the printed form is convenient for readers. It need not be consumed in real time, it is portable, and it is its own storage medium. These attributes form the touchstone against which electronic media must be judged.

Radio

The first important electronic mass medium was radio. The transmission medium was itself a public good, the first mass medium with this characteristic. This gave radio a tremendous economic advantage over print media in reaching large audiences for the sale of advertising, especially advertising with little information content. No print medium could reach so many consumers as cheaply. Radio had an additional advantage. Unlike newspapers, radio can cover real-time events, provided program origination facilities are present at the event.

In the early 1920s the radio industry was inchoate. There were a few experimental stations and a few thousand radio receivers. No one had clear rights to use the broadcast spectrum. No one knew what sorts of broadcasts might induce people to buy radio sets. Some potential manufacturers of radio sets imagined that they might need to supply free programming to stimulate demand. Others considered charging listeners an ongoing fee. The potential of advertising support was not readily apparent. Out of this chaos a few entrepreneurs made decisions and took risks that turned out to be highly profitable. One was David Sarnoff, who became head of RCA; another was William Paley, the founder of CBS. To bring order and predictability to their industry, and to control "excessive" competition, they promoted the idea of federal government ownership of the spectrum. They also invented networks. They discovered the enormous advantage of radio advertising over print.

Radio lacked certain useful aspects of the print medium. (Indeed, it is on account of these differentiating factors that print media have survived electronic competition.) Until the development of the transistor, radio was not portable. Before tape recorders, radio did not permit listeners to select the time when they would hear a program; radio was not a storage medium. Beyond these physical disabilities, the form of a radio transmission—chiefly the fact that consumption and production take place in real time—ruled out radio as a means of transmitting certain kinds of information, such as want ads. In addition, though there is no technical reason that a radio announcer could not convey the entire textual content of a newspaper, such a broadcast would be of little value to most listeners. The form of radio dictates less detailed and shorter messages.

Conventional Television

Television, on one level, is nothing but radio with pictures. It shares all the advantages and disadvantages of radio vis-à-vis print media. Of course, it turned out that pictures were highly valued by consumers. One reason is that television has a great deal of redundancy for any given form of content, which makes it easier to consume than radio. Long sequences of frames in a television broadcast tend to be identical or at least very similar, providing the viewer with an opportunity to dwell on the information, or in effect to have some control over the pace of consumption. In this respect television has some of the attributes of print that radio lacks. In sum, television does the same things that radio did, but much more expensively; therefore it could succeed only by adding new features that increased its value to consumers (or, in principle, to advertisers) in line with its added cost.

Cable Television

Cable television provides an interesting contrast. It was far from a revolutionary addition to the mass media family. Cable television provides exactly the same service as over-the-air television insofar as any given program is concerned. Indeed, it is hard for viewers to distinguish a cable program from a broadcast program. Cable television provides features superior to over-the-air television in only two respects: it has far more channels, and it permits viewers to vote with their dollars for the programs they prefer.

There is no technical reason why the number of over-the-air TV channels could not be as great as the number on cable, and there is no technical reason why stations could not charge viewers (by scrambling their signals, for example). Moreover, there probably is no economic reason either. The cost of building (around $1,000 per household) and running (around $15 per household per month) a cable system of given channel capacity in large markets certainly exceeds the cost of building and running a similar number of over-the-air broadcast stations and arranging to collect viewers' subscriptions to individual channels. If this is so, over-the-air television trumps cable television as a mass medium, and there is no technical or economic or product differentiation reason for the existence of cable television systems, at least in urban areas.

Why then is cable television an important industry in the United States and other countries? The answer is straightforward: the U.S. government (and foreign governments as well) prevented broadcasters from offering services that consumers wanted. Specifically, the American government, through its control of the radio spectrum, severely limited the number of broadcast stations. In addition, for many years it did not permit broadcasters to charge viewers directly, thus limiting the resources available for program development. By restricting output, the government acted as if it were a cartel manager, seeking to drive up the value of TV station licenses.

In a free market, one in which broadcasters could buy spectrum as needed to satisfy consumer demand for programming, there would have been no cable television industry. The same service would have been supplied more cheaply over the air.

Storage Media

Storage media, such as blank videocassettes, today are interesting chiefly as means to augment the television medium rather than as media in their own right. The videocassette brought viewers the ability to "time shift"—that is, to record programs for later viewing. This provided television with a feature that had previously distinguished it from print media. But time shifting was no more than a modest success. Television broadcasters had already adopted a mode of programming that minimized the disadvantages of having to consume television in real time. For most people, apparently, one television show is a pretty good substitute for another, and few shows are sufficiently better than the average to be worth the trouble of programming a VCR. Accordingly, most VCR use is limited to playing rented movies.

One could regard the physical distribution of prerecorded cassettes for sale or rent to consumers as a mass medium. Tape distribution has advantages and disadvantages vis-à-vis print and conventional television. Cassette distribution is at least as expensive as print distribution and far more costly per unit delivered than over-the-air broadcasts. The content of the tapes is, or could readily be, available over the air. Being able to charge consumers directly is an advantage, but only because of the artificial constraints faced by over-the-air television. Prerecorded tapes have relatively little value to advertisers because the timing and context of their consumption cannot be measured or controlled. Tapes do have the advantage of being a storage medium, which is clearly of value to those consumers who purchase rather than rent. In a rental context, the fact that tapes are somewhat durable is incidental. (The business would not be much different if a tape could be used only once and was manufactured to order in the store.) Durability, however, provides consumers with the opportunity to consume the product on their own schedule rather than in real time. The videotape rental "medium" exists largely on the same basis as cable: as a path by which consumers' demand for video entertainment can be satisfied despite regulations attempting to limit supply.

Common Ground

On the technical side, media have much in common. All transmission media, from newspapers to radio to the Internet, involve tradeoffs among *channel capacity (bandwidth), storage,* and *computing power (processing or compression).* Each combination of these three attributes has a unique form that conditions message content, determines creative scope, and underlies consumer preferences. Thus a given combination defines a particular medium. Efficient media reflect combinations of these attributes that minimize cost for any given level of information delivered to the consumer. Nonelectronic media have analogous attributes; for example, newspaper press capacity (pages per hour) is analogous to bandwidth.

Figure 2.1 illustrates tradeoffs among bandwidth, storage, and processing in producing a given data flow. Any point on this surface yields the same result from the user's perspective in terms of arriving information content. Higher or lower surfaces would correspond to higher or lower rates of data flow. Each point on the surface has a different cost. A successful technology is defined by that particular point that minimizes cost, given this technology set and the relative

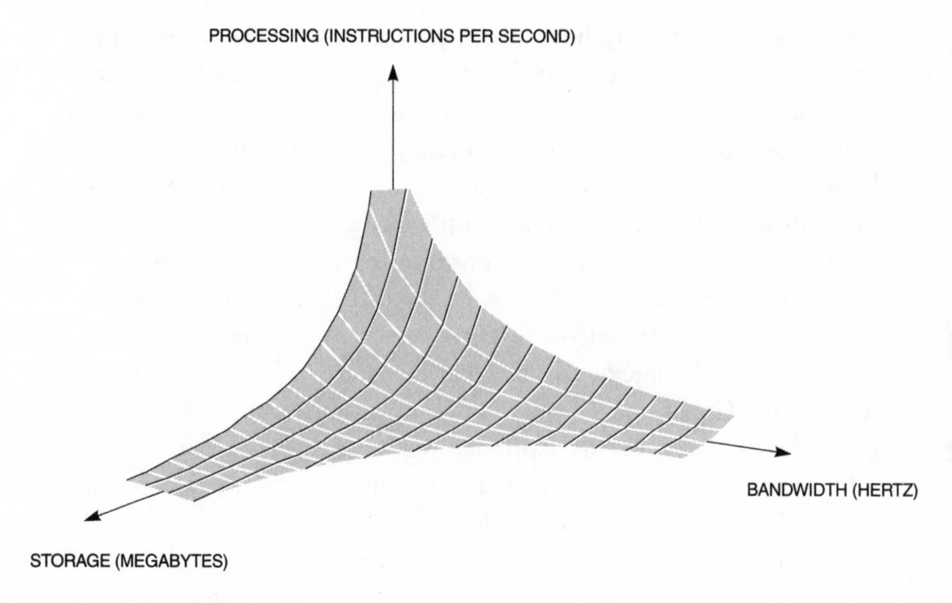

Figure 2.1 Bandwidth-storage-compression tradeoff

prices of the three components. Holding other inputs constant, there are diminishing returns to the application of any one input. Technological innovation is causing this surface to shift rapidly inward, but in unpredictable and uneven jerks. Meanwhile, the same thing is happening to prices: the prices of all three components are falling, but the relationship among the prices is constantly changing.

The term "bandwidth" is often used loosely to refer to the data-carrying capacity of a medium, rather than in its narrower sense as a measure of the range of frequencies used by the medium in its electronic or optical transmissions. Indeed, it has been popular in recent years to refer to the great "bandwidth shortage" caused by the explosive growth of the Internet and of video broadcast media. No shortage exists: there is a supply of data transmission capacity, with more forthcoming at increased prices. The method by which any given amount of capacity is produced will vary depending on the technical tradeoffs and relative prices of the ingredients of production. Chief among these ingredients are bandwidth (in the narrow sense), storage capacity, and processing power.

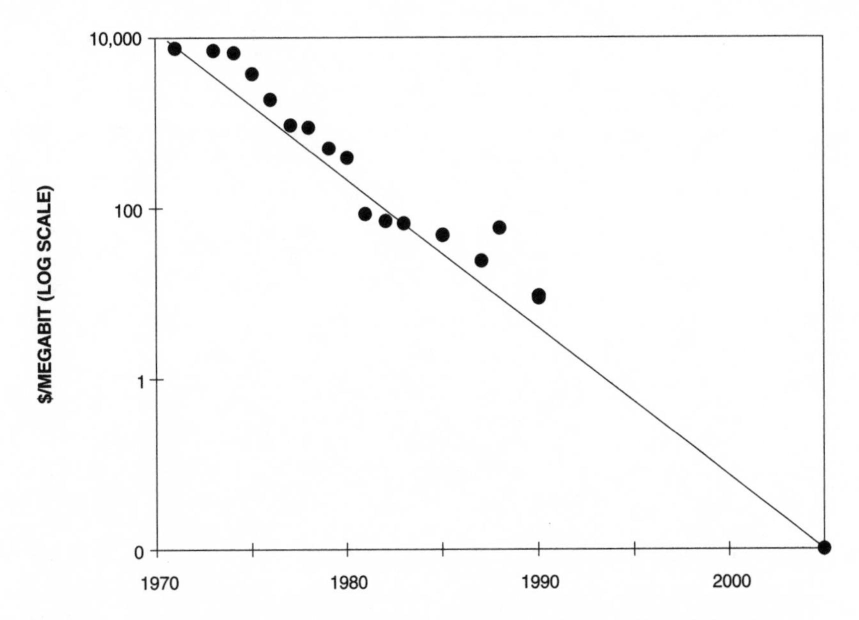

Figure 2.2 Cost of memory chips with projection, 1970–2005 (Technology Futures, Inc., 1995)

Figures 2.2 and 2.3 illustrate the declining unit cost and increasing performance of microelectronic devices and the increased availability of bandwidth. According to "Moore's Law," the unit cost of microelectronic devices declines by half every eighteen months or two years, while performance doubles. (This is an empirical, not a theoretical, observation. Industry observers claim it has held true since the debut of the ENIAC computer fifty years ago.) In part because of these falling costs and in part because of network effects (discussed later in this chapter) and improved interfaces, traffic on the Internet has doubled two to five times per year for more than fifteen years. Recent rapid changes in technology have

- introduced new digital media with far greater capacities in all three dimensions at much lower costs,
- changed the technical relationships among these three attributes in sudden hard-to-predict jumps, and
- shifted the relative prices of the three attributes.

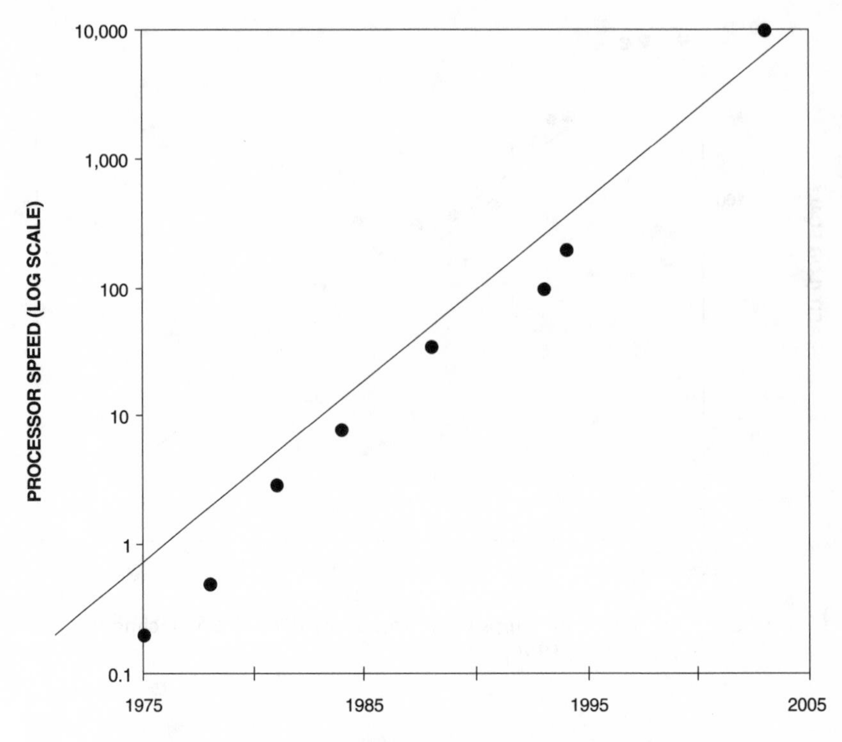

Figure 2.3 Illustration of Moore's Law: speeds of Intel processors in millions of instructions per second (Technology Futures, Inc., 1995)

In addition, the FCC has greatly expanded the radio spectrum available for wireless video transmission, which will result in falling prices for bandwidth. The increased bandwidth, however, comes with the constraints and risks associated with government regulation.

Consider Figure 2.4. The columns represent the total cost of any given quantum of communication, and the shaded areas represent the proportion of that cost attributable to bandwidth, storage, and processing. As time passes, technology, economies of scale, input price changes, and other factors cause the columns to get shorter. But the costs of the elements do not fall in lockstep with one another; they fall at different rates.

If we were to assume, for example, that wireless and wireline (telephone) media had the same costs and cost components today, we could be sure that in ten years both would cost less. But we have no

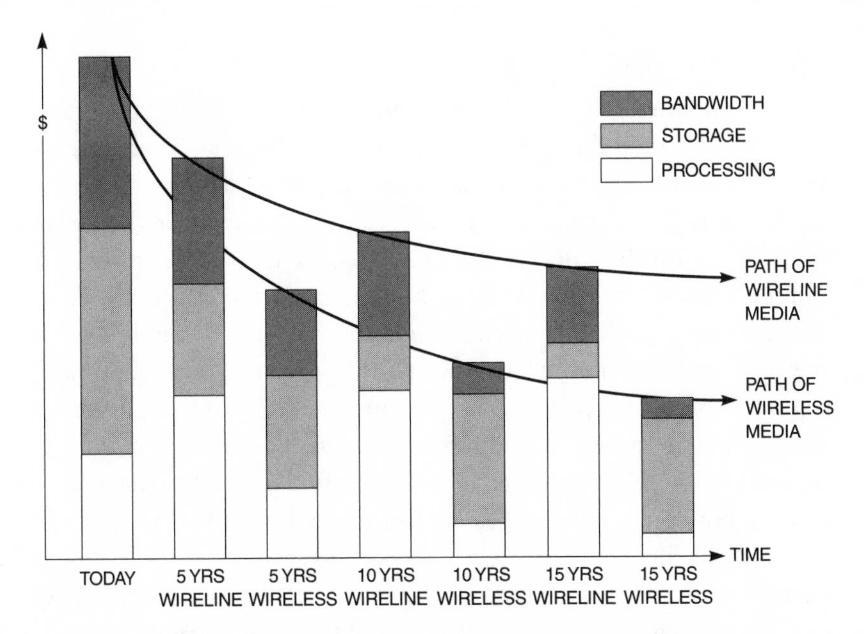

Figure 2.4 Costs and the competitive struggle between wireline and wireless media (hypothetical)

way of knowing which would be the cheaper medium, and we have no way of knowing which element(s) of cost would have been responsible for the victory of that medium. In Figure 2.4 it is assumed for purposes of illustration that wireless media win the race by taking advantage of a technology that relies heavily on bandwidth, in contrast to wired media, which (hypothetically) bet mistakenly on processing. The left-most column represents the (assumed) initially equal cost and cost components of wire and wireless communication at the outset. As time passes, wireless communication relies increasingly on bandwidth, producing a downward trajectory of overall cost that lies beneath the wire trajectory.

In order to avoid confusion, from this point forward "bandwidth" will be used, except where noted, in its narrow technical sense of a range of frequencies. The information transfer capacity of a given medium, arising from the medium's use of bandwidth (narrow sense), storage, and processing power, will be referred to as "capacity."

Form and Function

The structure or "topology" of a communication network also helps define the nature of the messages it can effectively carry. Three important distinctions are useful here. A broadcast, defined as a one-to-many, one-way communication path, can take place over the air, in print, or on the Internet. But if the object is to make a broadcast communication, then a network like the Internet is much closer to print than to over-the-air broadcasting. In print media, a public good (the message) is conveyed by means of a private good (the book, magazine, or newspaper). In over-the-air broadcasting, both message and medium are public, which typically conveys a substantial cost advantage. Internet "broadcasts" are exactly like print "broadcasts"—separate but identical messages are sent to each member of the audience. Finally, there is "interactive" video. This can take forms such as video programming called up on demand (VOD), or videoconferencing. The video message arrives in response to a request from the user; therefore, the message goes to only one user and is not broadcast. In this sense, a video rental store could be considered "interactive video." The Internet is interactive in a sense because all users can send and receive messages. The Internet, however, was not designed to be interactive in real time, and it does not ordinarily reserve an open channel to ensure prompt delivery of requested messages. Online services such as America Online once attempted to offer real-time features such as chat rooms by confining users to proprietary networks, whose performance can be controlled, rather than relying on the Internet, although chat facilities are available there as well. In short, the Internet has what is called a latency problem that makes it unsuitable for many video applications, whether interactive or broadcast.

The technology of any medium determines its suitability for broadcasting or for interactive video. Over-the-air television (whether terrestrial or from space) is an especially cheap way to transmit video broadcasts, because it takes advantage of the public-good character of transmissions. All of the advantages that digital technology brings to the Internet are also available to over-the-air broadcasters and satellite operators. A medium such as the Internet makes more sense (leaving latency aside) if the video delivery service is to be interactive. It makes particularly good sense if the senders as well as the receivers are widely scattered in space. While such a service can certainly be offered over the air (except for bandwidth it is functionally the same as a cell phone system), doing so would be quite expensive.

The "Internet" is a convenient word. It lumps together a disparate and changing collection of technologies and services accessed through personal or other computers, including those owned by businesses, governments, and universities. The range of Internet technology is broad. It encompasses more narrowly defined media, such as telephony, radio and television, and even newspapers, all of which now are supplied, in some form, over the Internet. The Internet can also deliver television pictures, but this does not mean that conventional (noninteractive) broadcast digital television will not coexist with it.

Any message traveling from one human to another at a distance must be impressed upon a medium (such as paper, magnetic tape, or carrier wave) by encoding (such as typesetting, digitizing, or modulating) to suit the features of the medium and then decoded or transformed back into terms accessible to human senses. For a long time electronic communication used analog modulation for this purpose, because it happened to be the most cost-effective way to get the job done. In recent years the cost of digital computers or processors has fallen greatly. Such computers can be used to lower the cost of communicating any given amount of information, if the information is in digital form, by conserving relatively expensive bandwidth. At the moment, digital coding is still more expensive than analog coding, but this disadvantage is overcome by the bandwidth savings.

The form of the Internet medium, and hence the form of its content, is determined (on the supply side) by a set of tradeoffs and constraints such as the tradeoff between bandwidth and compression. *All* these tradeoffs are changing in ways that are difficult to predict. Thus the form of electronic communication in the next decade (and in the next century) depends largely on whether the costs of data storage fall faster than the costs of data transmission, and whether the costs of data processing fall faster than either storage or transmission.

Prices and Predictions

If data storage prices fall more rapidly than the prices of other components (thanks to digital video disks, the successor to compact disks, and follow-on technologies), then entertainment, information, and images may be called up from local storage in pieces, as needed, at the demand of instructions sent over "narrow" transmission pipelines. This possibility is explored in the radio station example below.

If the price of processing power falls more rapidly than the prices of other components (it has been falling by 50 percent every two years), then the use of digital compression will further increase (it is already about 10:1 in some video applications and climbing). Faster processing will substitute for increased bandwidth and local storage. Digital compression and decompression take time, however, even with fast processing. They create delays in responding to users' commands. Such "latency" tends to undermine the quality of real-time video services. Put differently, an increase in computing power and hence compression may save bandwidth and storage costs, but it probably will not support interactive video services without increases in either bandwidth or storage.

Beethoven and Bandwidth

An understanding of the term "information" will help us appreciate the tradeoffs among bandwidth, compression, and storage. From the point of view of the consumer at whom a bitstream is directed, not all bits are news. (Bits are zeros and ones that are used, like Morse code, to send or store information.) Some bits convey content that is already known; these bits are redundant—they contain no information. For example, suppose a radio announcer says, "We will now broadcast Beethoven's Ninth Symphony." From the perspective of a listener who (a) believes this statement, (b) already owns the same recording of the Ninth, and (c) for whom listening to the radio and listening to the stereo are equally convenient, nothing is accomplished by the subsequent broadcast. The broadcast is a waste of electrons, or capacity, that could be better used for other purposes. At least from the point of view of that listener, the broadcast conveys no information.

Suppose all the radio stations in a city collaborated on the distribution to all listeners of compact disks containing the music that would be played on the radio in the next month. Then all but one or two of the stations could go off the air, and the remaining stations need only broadcast instructions (perhaps automatically to CD players) telling listeners which selection to play and when. In effect, only one or two broadcast stations would be needed for this highly "compressed" signal, consisting of instructions as to the order of play of the CDs, but yielding a sound functionally equivalent from the listener's point of view to traditional radio. (Of course, traditional stations would be required for real-time broadcasting.) The bandwidth presently reserved for the remaining radio stations could be used for cell phones or some other real-time function. There is nothing in this example that requires digital media; it could be done with present-day analog broadcasts.

In the example above there is a clear tradeoff between bandwidth and storage. Another way to look at what has happened is to say that the radio broadcasts have been greatly compressed into a signal consisting of commands that permit the receiver to *reconstruct* the full message. The reconstruction can take place by calling material up from storage. But if the signal is digital, it can also take place by means of computation. That is, the transmission may be a set of instructions that cause a digital signal processor (in effect, a computer) to "reconstruct" or create an image or a sound.

Imagine that at some point in the Ninth a solo violin sustains the note E for six seconds. That sound is digitized (sampled) and broadcast as a stream of identical bits lasting six seconds. The sound can be compressed by sending a very brief series of bits telling the digital signal processor at the receiving end to "play" an E for six seconds. These instructions can be sent in much less than one second. The five or more seconds saved can be used to send other information. Music (and TV images) contain large amounts of redundant data that can be compressed in such fashion. In this way, a channel that once carried only the Ninth could broadcast all nine symphonies at once. Thus nine radio stations can be replaced by one by means of data compression. Bandwidth can be conserved by storage or by compression or both.

Something analogous to this has been proposed in the real world. Right now, consumers can subscribe to HBO or a NVOD premium movie service on cable or satellite. For a fee, movies from afar can be transmitted and viewed on one's television. Electronics retailer Circuit City, according to the *Wall Street Journal* (September 8, 1997, p. B6), is backing a disposable digital video disk called Divx. Consumers will purchase the disk to enjoy one showing of a recorded movie. The disk can be reactivated electronically for reruns. That is, a very brief narrowband communication will remotely trigger the avalanche of bits that is contained on a Divx.

If It Isn't Profitable, It Won't Happen

Television exists to sell audiences to advertisers. Television also exists to provide entertainment that entices viewers, who may or may not be charged a fee. Like any medium, television is a collection of technologies, standards, and government regulations. The economics of television is the process by which producers, advertisers, and viewers, interacting with one another and with the technology, seek private advantage. Commercial television content is the outcome of this self-interested interaction.

Because broadcast television is a public good, there are very large economies of scale with respect to audience size. The bigger the audience, the lower the cost per viewer to make the program. Hence two programs with the same production costs and the same inherent attractiveness can have very different profits, depending on which has access to the larger audience. Because of these economies, the producer with the larger potential audience can profitably offer lower prices to viewers and advertisers. So, other things equal, media that reach larger potential audiences have substantial advantages over those with less reach. In the case of broadcast television, both the program content and the airwave broadcast have these public-good characteristics. Public-good advantages do not lead to monopoly, however, because programs are differentiated products, and broadcasters can seek specialized or niche markets.

In many markets an understanding of the supply and demand factors discussed above would be sufficient to make predictions. In ordinary markets supply and demand are linked by prices, which serve as impersonal signals to reconcile producer and consumer interests. Communication media markets are more complex, in part because of network effects.

Network Effects

Network effects (also called network externalities) link one consumer to another, making each consumer's demand for any given medium dependent on how many others have accepted it. Partly because of network effects, new media may not be successful solely on their merits. Superior media may be excluded from the market simply because some less worthy medium got there first and enjoys a first-mover advantage. Or a now-superior approach may have been extinguished because another approach seemed better at the time. (The latter failure is called path dependency. The example of electric cars, which were developed at the same time as gasoline-powered automobiles, is sometimes used in this context. Our transport infrastructure supports cars with gasoline engines and not cars with batteries, even though we might now agree, on account of air pollution, that electric cars would have been better all along.)

One of the most compelling metaphors for network effects is Gutenberg's invention of the printing press around 1450. The invention of printing involved several technological breakthroughs. The development of suitable ink and paper was no less important than movable type. Inexpensive paper supplies were required. People had to be literate for there to be a demand for books; books do not themselves create literacy. All these and many other chicken-and-egg complementary developments were required to bring mass literacy to Europe. Gutenberg's movable type gets too much credit. Despite the gradually declining price of books, Europe remained overwhelmingly illiterate until the nineteenth century, when the literacy rate finally rose above 50 percent. According to some historians, it was not movable type or cheap books that were chiefly responsible for the eventual spread of literacy; it was the ripening of social, cultural, religious, and political preconditions. Movable type, in other words, was a necessary but not sufficient condition for cheap books, and cheap books were a necessary but not sufficient condition for broad social change. It is difficult in the face of this history to take seriously the claim that cheap microprocessors are likely, by themselves, to generate a social revolution early in the next century, even in the culturally predisposed Western countries.

The success of any technology depends to a great extent on economic forces such as network effects, path dependency, first-mover advantages, and economies of scale. The very best ideas do not get implemented all the time, or even most of the time, or as soon as they could be. This notion is very worrisome to Internet entrepreneurs. But there is almost nothing that can be done about it. Government interventions in standard-setting tend to make matters worse, not better, because the government has no more clue to the right outcome than does the market, and because government "solutions" are much harder to change than market outcomes. In no case is this more apparent than in the government's sustained and successful effort to control broadcast technical standards, the effect of which almost certainly has been to keep TV technology a decade or more behind computer technology. Of course, to support its policies the government must control technical standards, because technology might otherwise offer the market additional ways around the controls. This is one reason why government regulations tend to grow more extensive and complex with time.

Network externalities exist when my use of a particular service is worth more to me because you use the same service. My fax machine, for example, would be worthless if no one else, or only a few people, had such machines. Subscribing to a telephone service provides consumer value only because others also subscribe, either to the same service or to an interconnected service. As the example of interconnected telephone systems suggests, network externalities do not necessarily imply that the service is a natural monopoly. Similarly, all fax machines need to be able to speak to one another, but they need not all be made by the same company.

Government: The 500-pound Gorilla

The future is not a question of technology and economics alone; there are other factors. The most important is government regulation, which despite recent reforms continues to pick commercial winners and losers, and to tax but protect winners. The government, once it lets a technology or an industry into the "club," seems to feel an obligation to help protect it from new competitors and new technology. Government can nonetheless derail the evolutionary process through cross-subsidies, encryption controls, and other policies. This lumbering beast has pushed many promising technologies off the tracks. But government also can stimulate ground-breaking new technologies through research subsidies and demonstration projects. Such subsidies played a big role in developing the Internet and commercial communication satellites.

Television and a large part of the Internet (especially telephone access lines and any radio-based transmissions) continue to be subject to regulation by the Federal Communications Commission and by the states. Therefore, the habits and vices of the commission (or, more generally and more accurately, the government) are going to be central to the development of new forms of video delivery. At this moment, telephone companies claim that the government is forcing them to subsidize the Internet by exempting Internet access providers from certain fees. If this is generally believed, sooner or later the public will want something in return for the subsidies.

Broadcasters face a particular problem in settling on a business plan for exploiting their new digital airwaves assignments, and they do so heavily constrained by the political baggage that goes with continued regulation. For example, it would be surprising if the broadcasters were not required to maintain some form of free over-the-air television service, and they may face significant restrictions on their ability to combine their airwave assignments into local units of efficient scale, in order to offer a service with bandwidth comparable to that of competing media such as satellites and wireless cable. Like cable, any new digital broadcast service will require substantial investments in consumer equipment. Traditionally, consumers must be persuaded to make this expenditure. But the pace of change now is so rapid, and the uncertainty over the surviving format so great, that it may be necessary for broadcasters (and other providers) to offer the equipment at heavily subsidized prices in order to induce consumers to subscribe. Of course, this raises the financial stakes enormously.

Although one can ask whether Internet hardware and software are *capable* of delivering television as we know it today, that is almost certainly the wrong question to ask. (The answer is yes, it can be done, although poorly and expensively.) The real issue is what *transformation* of the current video medium will make it most suitable for Internet transmission.

Interface Design

A key perspective on form and content is interface design—features governing how the user and the computer interact. The World Wide Web and the modern browser are chiefly responsible for the growth in demand for Internet services. Yet these revolutionary inventions added nothing whatever to the *functions* performed on or by the Internet; they merely made it easier to use: within a few years millions around the world began to log on. But the Internet is not yet a mass medium. Another leap forward into simplicity is needed. Indeed, advances in interface design may substitute for better transmission and computing hardware in defining a successful digital medium; lower costs mean nothing if the service has no value to users.

Cable companies, broadcasters, telephone companies, and others in business to deliver video entertainment to the home are paralyzed with indecision as they brood over investments running into the tens of billions. Investors are hesitating not merely because they don't know which technology will deliver conventional video channels most cheaply or because they don't know what the government may require or permit. They are stuck primarily because no one understands the best form in which to deliver the video. The answer to that question turns in great part on the hardware people will use to view television (or whatever it is going to be called) and the content consumers and advertisers will be willing to pay for.

While the hesitation of television investors goes on, the Internet does not wait. It grows and changes based on the supply and demand for nonvideo information and entertainment services aimed largely at an elite minority of the population. Transmission pipelines into the home increase in size slowly. Meanwhile, Internet commercial interests invent ever more clever ways to use limited bandwidth, substituting processing power, storage, and adaptive content. As they do so, the Internet becomes increasingly a substitute for television viewing rather than a means of television viewing. It also provides advertisers with substitute ways to reach viewers. The Internet has begun to pull viewers and advertisers away from television, and no doubt before long it will begin to bid away some of the creative talent that produces television content.

Entertainment Format

The Internet will not replace television; we will not fire up our personal computers to tune in *Seinfeld* reruns. Although the Internet and television share portions of the transmission network hardware, the computer will not replace the TV set as a viewing device. It makes no sense. Most of the things for which a computer is useful, including interactive entertainment, require the user to be next to the hardware. With broadcast television, it is the opposite. If what is imagined is the TV set as a Web browser, then we are envisioning not a personal computer/television but, rather, a television with a specialized digital capability—and one that seems very limited, since much Web use (for example, downloading applications or documents) requires a personal computer for implementation.

As for television itself, two things are clear: the government-created artificial scarcity of spectrum will cease to be a defining factor in the television industry, and the days when most viewers do not pay for most programs are numbered.

If, contrary to the tacit assumption of the present analysis, the price of bandwidth falls more rapidly than the prices of other components, the future of the Internet and television will depend on who owns rights to the technology. Unlike storage and processing, which are supplied by competitive decentralized industries, transmission facilities are often highly concentrated. The single most important (expensive) link in the transmission chain is the one that ends in the home. If it should turn out that the key transmission technology is controlled by local telephone companies, for example, the future will be very different (and much slower in arriving) than if one form or another of digital broadcasting turns out to be dominant. The digital assignments recently awarded to existing TV stations may provide an efficient broadcast path, but other possibilities include "wireless cable" (multichannel multipoint distribution systems, MMDS; local multipoint distribution systems, LMDS; and 38-gigahertz systems), geosynchronous communication satellites (GEOs), and low Earth orbit satellites (LEOs) such as the $8 billion Teledesic system that Bill Gates and Craig McCaw have started to construct.

Cost, the Biggest Unknown

After the problem of interface design and entertainment format, cost is the most important unknown factor affecting the future of digital media. The cost characteristics of these media break down into two categories: network costs and terminal equipment costs. Terminal equipment is the new box that sits on the television or next to the computer, and perhaps the antenna that goes with it. Network costs are those related to building and launching satellites, fiber-optic cables, or radio transmitters. Essentially all of the new video age scenarios involve rather expensive new terminal equipment, and there is enough uncertainty about the cost of this equipment to say that all the scenarios are about the same in this respect. Network cost characteristics, however, are very different. A geosynchronous direct broadcast satellite—or even a LEO system—has huge up-front costs but virtually no variable costs until 100 percent channel capacity utilization is reached. The same is true for the Internet backbone. Such networks have substantial fixed costs, shared by many users, which reduce the cost per user. At the other extreme is the telephone system with its star-shaped networks. Expanding the capacity of any subscriber line is just like adding terminal equipment, and the aggregate costs quickly become astronomical.

Satellite broadcasting (for example, DirecTV, Primestar) enjoys enormous public-good transmission advantages at the price of huge fixed costs. Rupert Murdoch's defunct GEO Sky project, with eight to ten satellites, was supposed to cost more than $5 billion (see Chapter 13). Provision for delivery of local programming was included in an attempt to make the Sky service comparable to cable and terrestrial broadcasting. All three could provide the same services at about the same level of quality (leaving aside the federal restrictions on broadcasting). Each of the 100 million U.S. television households would have had access to Sky. The space segment costs therefore amounted to $50 per household. More substantial are the costs of buying and installing receiving equipment and antennas. These costs are in the vicinity of $300 per household, bringing the total cost per household to $350 if all households were to subscribe. This would provide about 200 channels of television (similar to what is on current satellite systems) plus local TV stations for most viewers. While much lower than cable costs, this figure is more than double the cost of terrestrial broadcasting.

It follows that, assuming equal quality service, satellites should drive cable out of the market, but in turn be unable to survive in competition with conventional broadcasting. But quality is not equal. Satellite broadcasts are of remarkably high reception clarity, and major satellite systems can have more than 200 channels, including "near" video on demand. Cable can match this only by investing even more per subscriber in digital upgrades to its systems. Terrestrial broadcasters cannot match this by any means at present because of federal restrictions. Although future digital broadcast systems may present some useful opportunities, a capacity disadvantage will remain.

Using digital technology, telephone companies could supply TV signals over existing copper wires to households. A separate pair of wires would be required for each TV set in the home, but telephones and television could share lines. There is no serious limit to the number of channels that could be made available. Only one channel could be on the wire at a time, but the node serving each neighborhood could be supplied, via fiber, with a full complement of cable networks and "near" or even actual video-on-demand channels. Consumers would be able to select which channel came over the wire and to change that selection easily. The cost of doing all this is in the same range as the cost of installing and/or upgrading cable systems: roughly $1,500 per household, far more than the price tag for satellite broadcasting. Although the costs for telephone companies will fall, competitors have access to the same underlying technology, and their costs will also fall.

Currently deployed and future terrestrial multichannel wireless cable systems such as MMDS could offer the same quantity and quality of channels as satellites, but only by using expensive subscriber converter boxes (like those used for satellite broadcasts) to accommodate highly compressed video broadcasts. Assuming that these end-user costs are as low as those of satellite end-users, the issue is whether MMDS systems operating in every local area would cost more or less, in the aggregate, than a system that can reach all the same viewers from a single satellite. Given the wide range of uncertainty about all these cost figures, MMDS or similar high-channel-capacity terrestrial microwave broadcast systems cannot be ruled out as being competitive with satellites, and certainly not as being more economical than cable for equivalent quality. It is no doubt for this reason that some of the large regional telephone companies have been investing in these technologies.

What Lies Ahead

The arrival over the next few years of digital television sets will do little to separate the winners from the losers among media. The advantage of the sets is that they eliminate the need for most or all of the expensive electronics in the present set-top converter boxes. They probably do this, however, for all the competing media, so no single medium gets a particular boost from digital TV sales.

No one knows which if any of these video media will succeed, and the reason why no one knows is interesting: the different approaches all seem to cost about the same, and they all seem to offer features, and suffer handicaps, that are difficult to evaluate. Analysts with the latest engineering and marketing information can get out their spreadsheet programs and calculate away, but uncertainty about the assumptions overwhelms any attempt to conclude that one of these media clearly is superior to (more profitable than) the others, especially in light of changing technology and the effects on costs of network externalities. For example, a relatively small improvement in compression technology (such as digital subscriber lines) could make narrowband telephone (the present system) the long-term medium of choice for both Internet and even conventional video. No supplier wants to be in the middle of a multibillion-dollar investment project, only to have a small change in the technology make its efforts obsolete. So both the risks and the stakes remain enormous.

The immediate threat to conventional broadcast television is not the Internet. It is broadcast satellites and advanced wireless cable systems, all of which can use the latest digital compression technology without being constrained by government standards, and all of which have far more bandwidth than individual broadcasters do or will have. The threat from the Internet is not that Internet suppliers will be outbidding TV networks for Hollywood sitcoms, but that the frenetic Internet community will tempt broadcasting's mass audience over to the Web. If this happens it will be because the public has fallen for e-mail, chat rooms, multiplayer games, and other to-be-discovered Internet content.

Successful new media arise because they provide the same services more cheaply or new services that are more valuable to consumers. The Internet cannot provide television (as we know it today) more cheaply. Therefore, it can successfully challenge television only by offering a new kind of service that consumers find more valuable. No one has yet invented that service. Understanding the uncertainties faced by Internet entrepreneurs does help to explain one phenomenon: the hype.

There can hardly be an industry whose products, projects, strategies, and executives are better publicized than the Internet. The hype makes sense because it is in part through such means that the Silicon Valley herd can be stampeded in the direction of a particular medium, whether or not that approach is the best or the least costly. We can look forward to a lot more hyperbole.

Conclusion

Designing a successful medium involves complex tradeoffs between factors such as processing power, bandwidth, and storage. These elements are substitutes in transferring data. Predicting future developments is confounded by the fact that, although each element's cost is falling, costs fall at different relative rates at different times. The most cost-effective combination is therefore in constant flux. And there is no guarantee that the most cost-effective medium will prevail in the market.

Video delivery in the future is likely to be digital, but that does not necessarily mean that video will be sharing a common digital pipe with everything else.

It is not at all clear that the Internet is going to be an important mass medium of any kind, much less an important means of delivering television. It still lacks a user interface that can support mass market consumer access. Further, the Internet lacks the capacity to transmit television signals to mass audiences.

If the Internet does become a mass medium, it is more likely to be with nonvideo content than with video programming of the sort available on today's television media.

Development of the technical infrastructure required to deliver the next generation of digital video and Internet communication services requires investments so massive that competitors must "bet the company" when they choose a technology. There is great uncertainty as to which technology or technologies will be winners. Accordingly, despite the hype, few if any companies are actually making the required investments.

PART TWO

THE ANALOG COMMUNICATION WORLD

CHAPTER THREE

NONELECTRONIC MEDIA

Print, film, and other nonelectronic media have many of the same functions as the new electronic media, even though there are major differences in technology. They therefore serve as a useful frame of reference when heading out into the less charted waters of electronic communications. Most of the economic characteristics of electronic media are shared by print media and film: they are public goods, they experience network effects, and they enjoy economies of scale. Like electronic media, film and print media have the opportunity to control risks by such means as vertical integration and government regulation.

Print Media

The vertical structure of print publishing includes writing, editing, printing, distributing, and reading. These functions are common to newspaper publishing, book publishing, and magazine publishing, and they have obvious parallels in television broadcasting. Written material makes its way to readers through manufacturing and transport channels that have capacity constraints akin to bandwidth. Written material can be stored (as in a library) as an alternative to transportation. The format of the printed material reflects tradeoffs between readability and compactness. Finally, print media are supported by combinations of advertising revenue and subscribers' payments.

Unlike electronic media of all kinds, print media have not been subject to much government regulation. Government seldom, for example, imposes entry barriers on print media. On the other hand, print media, as a rule, are neither subsidized nor specially taxed; they have no "public service obligations" imposed by regulators. Aside from the environmental, health and safety, employment, and antitrust regulations that all businesses must confront, print media are not regulated. The First Amendment, of course, is chiefly responsible for this. Unlike the electronic media, print media can be wholly irresponsible, obstreperous, and innovative.

The Internet in form and content is a lot more like print than it is like television. The analogy of a newspaper has already been suggested as a way of understanding bandwidth and other features of the new media. Electronic media transmission capacity is achieved through various combinations of bandwidth, processing or coding, and storage. The written word is nothing but a code for information originally oral or intellectual in form. The coding can be concise or not, easy to read or not, illustrated or not; these possibilities correspond to degrees of data processing or compression. Newspapers and books are among the pipelines or channels by which this encoded information reaches us. Their distribution channels (presses, trucks, newsstands, bookstores, mailbags) correspond to the concept of channel capacity. Finally, the message of each of the print media is embodied in a physical object with varying degrees of permanence: a book, a newspaper, a magazine. These publications correspond to electronic storage media such as disk drives, caches, and CD-ROMs.

Another electronic media issue is standardization. Whether privately or through government regulation, many electronic media are required to conform to uniform technical standards that ensure, among other things, operability by numerous users. No official body establishes standards for "coding" English grammar and spelling (although there are such authorities for French and Spanish). No authority decides on standard sizes or formats for books, magazines, or newspapers, despite the obvious advantages that standardization would bring for libraries, printers, and those who transport printed media.

Does the print world suffer from this lack of standards? In some ways this question can be answered objectively. For example, it would be possible in principle to calculate the cost savings in the distribution and storage of printed matter that would arise from standardization of the sizes of books, magazines, and newspapers. It is difficult to say how much if any content would be excluded by standardization, but a great deal of content would have to change form and thereby (almost by definition) lose some of its value. Some publishers would face higher costs as a result of having to employ a less than optimal form for some or all of their publications. Ultimately whether the costs exceeded the benefits would depend, not on whether many publishers had to adopt a less-than-ideal form, but on how much the value of their product was impaired by that change.

The "system" organized around print media has strong network externalities. The existence of one author favors other authors because it increases the viability of printers, publishers, libraries, and bookstores. Similarly, each printer, publisher, library, and bookstore benefits from the existence of other entities of the same kind.

Despite strong network externalities, there is little vertical integration in the print media. Authors usually are independent agents without long-term employment agreements with their publishers. (Newspaper reporters are an exception.) Most publishers (except newspapers) do not own presses. Few retail distribution outlets (bookstores and newsstands) are owned by publishers, although again newspapers are an exception to the extent they engage in retail delivery service.

Newspapers are an exception for two reasons. First, they are published on a regular and demanding schedule, which leaves little time to negotiate commercial arrangements for each delivery cycle; this seems most obvious in the case of reporters. No newspaper publisher can rely on an ad hoc labor market to hire reporters as necessary to cover breaking stories. Second, and probably more important, daily newspapers tend to be the only local buyers and sellers of the services into which they have integrated. For example, in most cities the local newspaper publisher is the only potential customer for the services that a printing press of appropriate size can provide, or that reporters can provide. Therefore, the publisher as customer has a good deal of bargaining power, the consequences of which will lead to higher risks and therefore higher costs for those suppliers that do not have alternative customers. The same is true of distribution services. In these cases, vertical integration may well have more to do with minimizing bargaining and risk costs than with efforts to internalize network externalities. That is, but for the special bargaining and risk costs of selling services to local daily newspaper publishers, the publishers would be able to obtain such services at the same cost whether vertically integrated or not.

The entry of publishers into the digital age is a fascinating process to observe. Such entry has been most rapid for newspaper publishers, who have fewer concerns about protecting their intellectual property than book publishers have. (There is still no practical way for a publisher to prevent the copying and widespread distribution—pirating—of material "published" on the World Wide Web, but several technological fixes are in the works.) News content, because it quickly decreases in value, requires less protection.

According to the web page of the Newspaper Association of America, 722 daily newspapers had sites on the World Wide Web as of April 1998. This is about half of all dailies in the country. Many of these charge users for "premium services," such as archival materials or, in the case of the *New York Times,* the daily crossword puzzle. The *Wall Street Journal,* with 70,000 Internet subscribers, seems to be alone in charging a subscription fee for basic access to its site.

Most on-line newspapers sell advertising, which appears in small windows or banners near the textual matter. The ads are often colorful, with animation or streaming text, and they frequently introduce significant delays in downloading pages. Among the unanswered questions: How effective are such ads? How can viewers' exposure to the ads be measured? How much should advertisers be charged? How much, if anything, should users be charged for access or for features? Will sales of the printed version of the newspaper suffer unduly from electronic sales? As to the last, the printed versions seem safe for a while. It is difficult to lug a computer, modem, and monitor to the breakfast table to read the paper over toast, or aboard a train for the ride to work. Even a laptop computer with a wireless modem is less convenient than a newspaper.

The Internet cannot fully displace print media until or unless a display device as convenient as a book, magazine, or newspaper is developed. Even when that happens, the services of authors, editors, and publishers will continue to be required in the electronic environment because of their value to users. One of the many innovations that transformed the small, eighteenth-century newspaper publisher, operating a hand-powered letterpress, into a big business a century later was newsprint: cheap pulp paper in place of expensive rag paper. Even though this and complementary changes in typesetting and printing made it possible for William Randolph Hearst and Joseph Pulitzer to revolutionize newspaper publishing, the functions performed by publishers or the readers' needs served by newspapers changed little. Internet publishing could constitute a similar change. Readers will still demand the services of editors to protect them from the onslaught of information. Enterprises recognizably like publishers will be required in order to undertake the economic organization of news gathering, editing, and distribution.

The factor most affected by Internet delivery of newspapers is, obviously, delivery costs. Daily newspapers have increasing costs as the distance between printing plant and subscriber increases, which is one reason why we have local rather than solely national newspapers. Internet delivery cost is insensitive to distance. One result of widespread Internet delivery of newspapers may thus be a movement away from local and toward national and international publishing, and also toward interest group orientations. Minority points of view that are not geographically concentrated are costly to serve with traditional newspapers; Internet delivery makes these groups much less expensive to reach.

Film

The motion picture industry gives us another touchstone for understanding the new electronic media. Like print, film has public-good characteristics, economies of scale in distribution, and network externalities. Like print, film is produced in vertical stages:

- content creation (screen writing, directing, filming, acting, editing),
- risk bearing and editing (producers, studios, distributors), and
- distribution (theaters, television networks, videocassettes).

In this list of functions, "editing" stands for the function performed by studios, analogous to that performed by book publishers, of selecting the package of material (portfolio of films) that will be made available to the public.

Although many technological innovations have overtaken the film industry, these basic functions have remained. The new technologies are adapted because they increase the demand for one or more of the functions, or because they decrease costs. The advent of sound and color, for example, enhanced demand. The use of computers in editing reduces costs.

Films contain an enormous amount of information, and even more data, in the technical senses of those words. It takes around 100 million bits per second to transmit an uncompressed high-definition video corresponding in quality to a motion picture film. Neither information nor data, of course, have much to do with meaning. An action color film such as *Back to the Future* contains far more data and information than a black-and-white film such as *Casablanca,* but many critics would probably argue that the latter conveys more meaning.

Methods of distribution or transmission for films have changed a great deal, partly in response to technological progress and partly in response to changing tastes and consumers' needs. From the beginning there have been motion picture theaters, but the huge, single-auditorium downtown palaces have given way to suburban multiscreen operations analogous to multichannel electronic media such as cable television. The outdoor drive-in theater rose and fell with changes in Americans' use of automobiles and adoption of newer technologies such as the VCR. Premium cable television networks and then prerecorded rental motion picture cassettes sprang up overnight as distribution businesses, and they may disappear just as quickly if broadband transmission offers a cheaper or more attractive distribution alternative.

The motion picture industry for many years has exploited the economic value of its products by releasing films in multiple geographic markets and in multiple temporal windows. For example, new films are first released to theaters, where moviegoers are prepared to pay the highest prices. Later the films are released to rental stores, premium cable networks, broadcast networks, and television syndication. Each window garners viewers who are willing to pay less, on average, than those who purchased in the previous window. New distribution technologies have been successful in those cases where they present motion picture distributors with opportunities to establish new windows, or to make the old ones more efficient.

Déjà Vu

Pre-electronic and electronic media have much in common. For our purposes what is most important is that pre-electronic media set the stage for understanding the functions and economic characteristics of current and future electronic media. Successful media such as print and film take advantage of their public-goods character, of economies of scale in distribution, and of network externalities. Their ability to do so rests in part on vertical integration, risk bearing, and other familiar business strategies. When we look at television and the Internet, we should not be surprised to see these same phenomena.

Conclusion

Thinking about the economic and technical structure of nonelectronic media such as print and film, even as briefly as this, can help us understand and evaluate possible futures for the new digital media. Both print and film overcame such problems as network externalities; in general the solutions did not require monopoly control or extensive vertical integration. In a world of technological change each surviving medium must offer either a differentiated service or a lower cost of production.

CHAPTER FOUR

THE EVOLUTION OF BROADCAST RADIO

Radio and the Internet in one way have already converged. More than four thousand radio stations were broadcasting on the Internet by April 1998, reaching audiences far beyond their over-the-air reaches. In May 1997 audio programs broadcast over the Internet first became part of the Museum of Television and Radio in New York City. The five programs included one live broadcast of a seder from Temple Emanu-El in New York, and the other four were interviews carried by "GRIT," an "Internet radio broadcaster" (*CyberTimes,* May 17, 1997).

The very early history of radio serves in many ways as a paradigm for the current world of digital technology. In the decade or so before Congress passed the Radio Act of 1927, the radio industry progressed from an inchoate cloud of promising technologies with no clear commercial purpose and no leading economic actor in a position to exploit them to a purposeful and profitable commercial enterprise. For this to come about, the radio pioneers had to overcome many of the technical, economic, and political problems that face Internet entrepreneurs today. The strategies of the radio pioneers remain highly relevant.

Although history is not required to repeat itself, some patterns, especially those related to government regulation, could occur again. Further, the economic and technical basis for radio broadcasting sheds light on the future evolution of television and the Internet. Finally, it is instructive to note that while in some respects television may have "replaced" radio, radio in close to its original form is still with us, larger than ever (if smaller than it would have been without television). To be sanguine about the future of the Internet as a video medium is not necessarily to condemn television in its present form to oblivion. Radio survived the advent of television by finding new niches. One, of course, was service to mobile listeners such as those in automobiles. Another strategy was quite remarkable: radio literally receded into the background. In the 1930s and 1940s listeners paid attention to radio broadcasts at least as much as viewers pay attention to television today. Listening to the radio today is hardly ever anyone's "primary activity."

Order out of Chaos

At the end of World War I, radio was conceived as a military tool, as a means of "wireless telegraphy," and as a toy of amateur electricians. A number of relevant and conflicting patents were held by companies such as the American Telephone and Telegraph Company (AT&T), Westinghouse, American Marconi (the U.S. branch of the famous Italian company), and even United Fruit. (United Fruit relied on the wireless to link its Central American plantations with transport and other facilities.) There was no commercial broadcasting. There were no standards governing radio frequency assignments, power levels, or antenna heights. However, the Department of Commerce under the Radio Act of 1912 (passed partly in response to the *Titanic* disaster) did act to issue licenses and to prevent interference among stations.

We know from subsequent events that the public in 1920 had a latent demand for radio programs and for receiving sets, and that advertisers had a demand for audiences. Demand was sufficient in 1920 to sustain a profitable broadcasting industry, just as it is today. Of course, this was not obvious to anyone in 1920. No one was sure whether there would be significant audiences for radio broadcasts. Many were convinced that advertising support ("sponsorship") would not work as a major revenue source. It was not clear whether radio set manufacturers would have to sponsor free programs in order to sell sets. The possibility of selling programs to listeners (subscription radio) had not been explored. As of 1920 there was exactly one radio frequency assigned for commercial broadcasting. All stations, separated either by distance or in time, had to share that frequency. And if radio should prove commercially successful, there was no way that new entrants could be prevented from flooding into the market to share in the booty.

Lack of information about the structure of demand for radio made investments in radio stations and in radio set manufacturing facilities very risky. Among other things, there was a "chicken and egg" problem. Consumers would not pay for radios if there were no programs, and broadcasters had no reason to transmit programs if there were no radios to receive them. Clearly, the more radios there were in American homes, the less risky it was to start a broadcast station. Not surprisingly, among the first broadcast stations were those owned by radio set and radio broadcast equipment manufacturers such as the Radio Corporation of America (RCA), General Electric (GE), and AT&T.

Even such fundamental issues as transmitter power and frequency bands were ill defined. For example, radio sets could be built to receive faint signals and to filter out noise, but that made the sets more expensive. One alternative was to make radio transmitters more powerful. Another was to increase the geographic spacing between stations to reduce interference. A third, after more frequencies were made available, was to increase the spacing between the frequency of each station. These choices involved engineering tradeoffs that would have been resolved routinely in a way that minimized overall cost if all radios and all stations were owned by one entity. But in the real world, stations and sets were manufactured and owned by (somewhat) different entities.

Commercial radio began, in part, as a publicity stunt for department stores. In 1912, the man who would become the giant of the commercial radio industry as chairman of RCA, David Sarnoff, was a lowly radiotelegraph operator assigned to a Marconi wireless station atop the Wannamaker building in New York City. Radiotelegraphy aboard ship was new; many vessels were not yet equipped with wireless equipment and others did not maintain round-the-clock watches in the radio room. The *Titanic*'s early distress calls went unheard by nearby ships, either because no one was listening or (in some versions of the story) because amateur radio operators drowned out the signals. The Italian aristocrat, inventor and entrepreneur Guglielmo Marconi, was present in New York at the time. He manipulated news of the disaster to promote sales of radio equipment and to persuade Congress to require wireless equipment and twenty-four-hour radio watches aboard ships.

Marconi (and later RCA) for promotional reasons chose to create the myth that Sarnoff had heard the *Titanic*'s distress calls and had then spent more than ninety hours at his post transcribing news of the disaster (especially lists of survivors). President Taft, in one version of this story, ordered all other wireless stations in the United States off the air so that Sarnoff could better hear the signals. Apparently none of this is true. But true or not it is an indelible part of radio history, and it played an important role in establishing both government and public attitudes toward radio, starting with the Radio Act of 1912.

As David Sarnoff understood, real time, immediacy, was radio's great selling point. Consumption of real-time events enhances the value of the medium because of the satisfaction consumers derive from shared experiences. The *Titanic* coverage was by radiotelegraph, not broadcasting. But by the thirties, everyone could "witness" real-time disasters, notably the *Hindenburg* inferno of 1937.

Sarnoff dominated the broadcasting industry, both radio and television, for three decades. In 1915 he wrote a famous "radio music box" memorandum, foretelling the commercial broadcasting industry. It was his idea to offer free programs in order to induce sales of RCA radio receiving sets. Later, he and William Paley, the founder of Columbia Broadcasting System (CBS), helped persuade advertisers that radio could be an even more effective selling tool than newspapers and magazines with their relatively small circulations. Sarnoff and Paley showed them that network radio could reach tens of millions, and thus become not merely effective as a marketing tool but extraordinarily cheap as well, measured on a per-listener basis.

Before the advertisers could be convinced, ways had to be found to reduce the economic or commercial risks of radio broadcasting. The first source of risk was technological. Who owned the rights to the new technology and was therefore in a position to dictate its direction? And what would happen if one firm went ahead with manufacturing that later was held to infringe another's patents? The second risk was associated with standardization. What frequencies would be used, with what separations between stations, at what power levels? And how could the evolution of these standards be managed? The third risk was competitive. Assuming every other uncertainty was overcome, and that a commercially successful business emerged, what was to prevent new stations from entering the market in competition with the pioneers, free-riding on their technical accomplishments, and driving down profits?

David Sarnoff at American Marconi and executives at AT&T, GE, and Westinghouse addressed these problems in an extremely effective fashion. First, regarding the technology, the corporations that owned the important patents formed a patent pool. The background for this was the formation of RCA, of which Sarnoff eventually became president. The United States government during World War I had relied on American Marconi and other foreign-owned radiotelegraph companies. Secretary of the Navy Josephus Daniels grew restive about the Navy's dependence on foreign suppliers and readily fell in with a plan put forward by Owen Young, the general counsel and later chairman of General Electric, to acquire Marconi's United States assets for $3.5 million. Then, to resolve the patent issues, Sarnoff distributed stock in the newly formed RCA to Westinghouse, AT&T, and United Fruit in return for their patent rights, thus forming a patent pool among the equipment manufacturers. By 1921, RCA was owned 30 percent by General Electric, 21 percent by Westinghouse, 10 percent by AT&T, and the rest by United Fruit and the public. Though potentially anticompetitive, the patent pool did eliminate the uncertainty and litigation costs attending the conflicting patent claims, thus promoting investment by members of the pool.

One important element of the patent pool was an agreement to divide the market. It was settled that AT&T would manufacture radio transmitters, while the two broadcasting companies, GE and Westinghouse, would manufacture receivers, with GE to have 60 percent and Westinghouse 40 percent of the receiver market. RCA would continue Marconi's radiotelegraph business and would be the sole sales agent for the receivers manufactured by GE and Westinghouse. The security of this "Radio Trust" was buttressed by its many patents, which made it difficult for any other firm to break into these markets. Technological uncertainty was thus greatly reduced or eliminated.

Not long content to limit RCA to distributing radio receivers, Sarnoff soon moved into broadcasting in competition with GE and Westinghouse, RCA's major stockholders. In 1922, AT&T also entered broadcasting with station WEAF in New York and sold its stock in RCA. AT&T modeled its business plan for radio on its telephone business: as a common carrier, it proposed to lease time on the station to anyone willing to pay the tariff. Moreover, it leased time on its telephone lines to those who wished to buy radio station time in distant cities, thus forming the first networks. Thereafter, when RCA, GE, and Westinghouse jointly attempted to enter the radio network business, AT&T refused to permit them to use its monopoly intercity telephone lines. AT&T found itself at odds with all the major companies of the radio industry, expensive litigation threatened, and by 1926 it decided to exit the business, selling most of its radio assets to RCA.

At about this time, Sarnoff managed to acquire a manufacturing subsidiary (Victor) and eventually ownership for RCA of all of the stock of its broadcasting unit, the National Broadcasting Company (NBC) and Victor. When the Justice Department finally challenged the Radio Trust for violations of the Sherman Antitrust Act, Sarnoff settled the case, in 1932, by having GE and Westinghouse spin off all their RCA stock. Sarnoff thus achieved autonomy at RCA, but the three firms thereafter competed in a stabilized market for radio equipment manufacturing.

Herbert Hoover Rescues David Sarnoff

To deal with the remaining sources of risk (lack of standards, competition, and new entrants), Sarnoff and his colleagues in the radio industry enlisted the help of Herbert Hoover, then secretary of the Commerce Department. Hoover thought it entirely appropriate that the government should assist the development of this new industry. Under the 1912 Radio Act, Hoover already had the authority to issue radio broadcast licenses in a way that minimized airwaves interference caused by more than one party attempting to broadcast on the same frequency at the same time and place. In addition, Hoover sponsored annual radio conferences at which industry representatives worked out differences and established standards, which were then implemented by the Commerce Department. As a result, between the establishment of KDKA, the first commercial broadcast station, in 1920 and 1926, millions of radio sets were sold and nearly six hundred radio broadcasters came on the air.

The burgeoning radio industry was cast into disarray in April 1926 when a federal court in the *Zenith* case decided that Hoover had no power to deny a radio license to an applicant on account of interference problems. Although he probably had other means to deal with the problem, Hoover decided to stand back and let the inevitable happen, and the industry quickly fell into "chaos" as interference among broadcasters increased greatly. Hoover apparently used, and may even have engineered, this circumstance to pressure Congress to pass the Radio Act of 1927. The act created a regulatory commission (the Federal Radio Commission, a precursor of the Federal Communications Commission) with the power to issue broadcast licenses. Among the first acts of the new commission was to reject proposals to expand the number of stations. The number of commercial radio stations remained below 1,000 from 1926 until after World War II, when competition from television made it pointless to continue to restrict output in radio. Today there are more than 10,000 commercial radio stations (see Figure 4.1). Herbert Hoover, having been the willing handmaiden, midwife, or pimp, depending on one's perspective, of the nascent radio broadcasting business, did not lack for recompense. His successful 1928 presidential campaign was the first to rely heavily on the new medium to reach voters.

In the 1927 act Congress in effect nationalized the commercial broadcast spectrum. Earlier licenses issued by the secretary of commerce were nonexclusive, like a ham radio license, and did not, for example, prevent nearby stations from broadcasting on the same frequency. (In fact, as noted above, there was at first only one frequency on which they *could* broadcast.) In practice, broadcasters worked with the Commerce Department to find ways to share frequencies by geographic separation and by time sharing. Under the new regime, the government for the first time had unlimited power to restrict entry into broadcasting, and it used this power to reduce the threat of competition to established broadcasters, thus encouraging their further development.

Congress had many motives in nationalizing the spectrum, among them the desire to prevent the development of private property rights in the airwaves. Sarnoff at RCA, like others in the industry, was a proponent of licensing because it removed the clutter of amateur broadcasters and stabilized the airwaves for commercial exploitation. By reducing the threat of competition, licensing reduced RCA's business risks. Sarnoff saw an advantage in being able to turn to the government for assistance, when necessary, rather than relying on the market and private property rights. Ever since passage of the Radio Act of 1927, the airwaves have been characterized as a "scarce" and "valuable" resource that belongs to the public, and any private use has been regulated to ensure promotion of that public interest. Similarly, the promotion of the Internet by government policymakers as a solution to various social and educational problems eventually may lay the groundwork for increased regulation and the elimination of competition that threatens these political goals.

The Radio Commission, and after passage of the Communications Act of 1934, the FCC, had to make two fundamental decisions. How much of the airwaves should be devoted to commercial broadcasting in each city? Who should be awarded licenses? As current experience indicates, the FCC could have chosen to license many radio broadcasters; instead, it licensed only a few in any given city.

Radio listeners in urban areas today have several dozen stations to choose from. In the 1930s and 1940s, however, most listeners had only three or four meaningful choices of local commercial radio stations. The relative scarcity of options was not due to technological limitations, and probably not due to insufficient demand from listeners or advertisers. It was the policies of the government that caused the scarcity. Since 1927 it had chosen to limit the number of competitors, initially perhaps to reduce risks that might discourage development of the industry, and later to preserve for incumbent broadcasters the economic rents created by this policy. This policy—the creation or promotion of artificial scarcity—taken both as a policy model and as the basis of important interest groups, led to similar policies with regard to television. It was not until mid-century, when television eliminated most of the economic rents in AM radio broadcasting, that the government permitted the number of AM stations to increase, as shown in Figure 4.1.

There are two explanations of why the government did what it did with radio in 1927. The "official" explanation was elaborated two decades later by Justice Felix Frankfurter in a Supreme Court case challenging the FCC's chain broadcasting rules (National Broadcasting Company, Inc. v. United States, 319 U.S. 190 [1943]). The rules restricted networks' ability to control the actions of their affiliates. Frankfurter made the following two points:

1. The commission limited the number of competing stations because of the scarcity of the airwaves (there simply were not enough frequencies to go around) and to stop chaos from interference among stations.
2. The commission restricted competition in order to ensure that its radio licensees had the economic wherewithal to serve their communities with adequate public interest or "sustaining" programming.

But it is an alternative explanation, an economic one, that best explains the historical events.

1. The commission restricted competition because it has always been susceptible to pressure from economic and political interests, particularly those of members of Congress and of incumbent broadcast licensees.
2. The commission consistently has reacted by protecting incumbent licensees from undue increases in competition. This made incumbent broadcast licenses ever more valuable and created ever-greater political and economic pressure to maintain the status quo.

On this political and policy foundation, television debuted at the 1939 World's Fair. Again, as with radio, it was necessary to decide how much of the airwaves to devote to television, and to whom the licenses should be awarded. It will come as no surprise that the an-

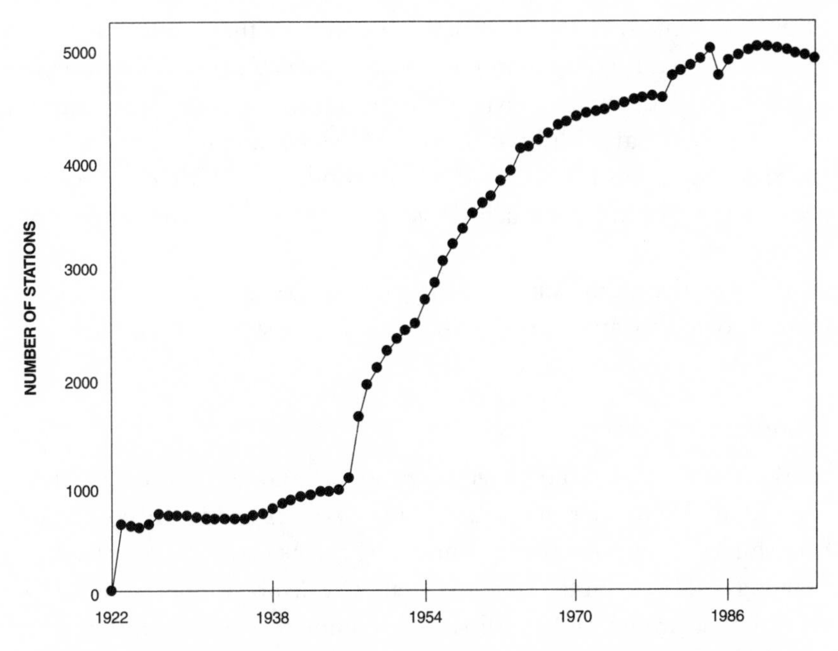

Figure 4.1 Growth of on-air AM radio stations (TV & Cable Factbook, 1997)

swers to emerge after the war were: *not enough spectrum to satisfy the demand* and *award licenses to the same folks who owned the radio licenses, whose value would be reduced by the new medium.*

David Sarnoff and the other radio pioneers were able to bring a highly profitable radio industry into being out of the confusion and uncertainty that characterized the early 1920s because they (1) internalized the problem of radio broadcast–radio set interdependence, in the early days, by having set manufacturers provide programming; (2) understood and promoted the use of advertising to support programming; (3) understood and developed radio networks (or "chains" as they were called in those days) as devices to lower per-listener program costs and to increase advertising revenue; (4) organized competitors to cooperate rather than compete in critical dimensions such as patent pooling and technical standards; and, probably most important, (5) enlisted the federal government as the industry's guarantor of protection from "excessive" competition.

This litany of accomplishments challenges our modern policy prejudices in two respects: the use of industry cooperation and the use of government to reduce competition. Both of these, especially the second, can reduce output and make consumers worse off. In the case of the radio industry, however, these restraints on competition arguably were necessary. Without them the business might have remained too risky to support much investment, resulting in little or no radio broadcasting, or at least much slower growth, again at the expense of consumers.

Many of the radio industry's early challenges, and the solutions to them, provide a framework for analysis of television and the Internet.

Economic Incentives

As the example of radio suggests, products and services do not become available to consumers merely because a clever scientist or engineer thinks of a new way that something can be done. Almost twenty years passed between the time Marconi first demonstrated transatlantic radiotelegraphy and the first U.S. commercial radio station was established. In a market economy, goods and services are available because someone expects to make a profit and is therefore prepared to pay the costs and to undertake the risks of production. New technology, however exciting, cannot come to market if it does not produce, or at least promise to produce, profits.

The very development of technology itself is responsive to economic incentives. Research and development are often very expensive. Not every idea can be developed. Generally, only the ideas that seem most likely to be profitable *are* developed.

Finally, every consumer has a "willingness to pay" (which may be zero) for every product or service. A particular good or service—however innovative, educational, culturally uplifting, aesthetically acclaimed, or otherwise meritorious—will not survive in a commercial setting if its cost exceeds consumers' willingness to pay.

Economics is a tool that can be used to explain or predict human behavior on the basis of certain fundamental assumptions. These assumptions are the "rules" of the game. For example, one can explain quite a lot about human behavior by assuming that people pursue their own economic self-interest, that they are rational, that they operate in a world where data are costly, and that they prefer less risk to more. While economics cannot say, except by fatuous assumption, whether any given consumer is likely to prefer apples to oranges, it can predict that if the price of fruit falls, consumers will buy more fruit.

Building on such assumptions, one can describe their implications for market outcomes, even government-regulated market outcomes. The issue is not whether people are, or ought to be, self-interested; the issue is whether the assumption that they are (or the assumption that they act *as if* they are) is a useful way to make accurate predictions about their observable behavior. As it happens, this assumption is a powerful predictor of human behavior. One can explain, for example, much of the behavior of David Sarnoff on behalf of RCA on the basis of such an assumption.

Private Needs and Public Goods

If a restaurant sells me a nice dinner, and I consume it, there is nothing left. No one else can consume the same meal. To serve two people, the restaurant must cook twice as much food and provide twice as much wait service as it does for me alone. What I eat and the service I get comes at the expense of someone else, who cannot eat or be served that meal. Remarkably, none of this is true for conventional radio broadcasting. If a station broadcasts one radio program, it remains available for others; it is not diminished by my consumption of it. It costs the producer nothing more to record a show for two people or a thousand people than for one person.

Economists call goods with these characteristics "public goods" because such goods, it was once thought, could not be commercially produced. But this proved wrong. Radio broadcasts are among many examples of public goods produced in the private sector. The private supply of public goods is possible when there are practical means to exclude nonpaying consumers.

Most nonbroadcast media have some elements of public goods. When one buys recorded music on a compact disk (CD), the content of the CD is a public good, but the physical vehicle for its delivery, the CD, is a private good. The musical composition itself is a shared resource, and no matter how many consumers share it, its value is not diminished. Similarly, the content of interactive television is a public good. Interactive television includes "video on demand" and "pay-per-view" television. Interactive television has public-good characteristics in its content (even though nonpayers can be excluded from that content), but it loses these characteristics when it comes to transmission. Each user requires a separate transmission, and these transmissions are private goods.

Who Pays and Why: Transaction Costs

Until recently, most radio and television in Europe was provided by government agencies. In an extensive early experiment with delivery of radio signals by wire, the British Broadcasting Corporation (BBC) relied on payments from listeners thus wired as well as on government subsidies, and not at all on advertising (Coase, 1950).

The model of government-sponsored broadcasting could have been adopted in the United States, but it was limited here to certain frequency reservations for educational broadcasters and later for public television. Notwithstanding the American political tradition, the novelty of radio broadcasting might have led Congress to provide for governmental (or at least educational) exclusivity for the airwaves. Those, such as Sarnoff, who were in control of events saw little opportunity for profit in such a course. That this possibility even existed, however, illustrates the substantial noncommercial or even anticommercial thread in United States broadcast policy.

Advertisers rather than listeners ended up supporting radio in the United States because of high transaction costs in the early days. Through the regulatory process, listeners developed equity rights in this status quo. Gradually, it became impossible to charge consumers for "free" over-the-air broadcasts. Transaction costs are the costs that buyers and sellers face in trying to trade goods or services for money. For example, if merchants face substantial losses from bad checks, they will insist on cash. Both bad checks and the need to carry cash to buy goods increase the cost of making transactions and thereby diminish their number and value.

Transaction costs are especially important for broadcast radio and television, when consumers are paying. In the early years of radio, transaction costs for collecting money from listeners over the air (in contrast to the BBC's wires) were so high that no one considered it; all radio was supported by advertising. The same was true in the beginning for television. This unremarkable, expedient business decision of the early broadcasters has, curiously, given rise to present-day claims that the public has a "right" to free over-the-air television broadcasts. Any number of politicians will defend this right, without regard to the costs imposed on the industry or on consumers. (Imagine a law requiring publishers to give away newspapers and magazines!)

An interesting unstated assumption of the broadcast business plan is that charging listeners directly for programming would be less profitable than offering it for free and relying on either advertising or equipment sales. As the BBC efforts with wired radio suggest, there was no fundamental technical impediment to pay radio. Probably pay radio could have been provided even without wires. But the pioneers took a different route. They may have thought that listeners would not be willing to pay (although that supposition is contradicted by the money spent on receiving sets); they may have thought that the transaction costs of collecting the money from listeners would make the effort uneconomic; they may have thought that the government would not permit the public to be charged for using the public airwaves. But the most likely explanation is this: they thought that if they charged listeners a sum sufficient both to cover the transaction costs and to make a contribution to the enterprise, fewer listeners would be available to sell to advertisers. They had to guess which value would be higher, the value a listener would put on programs (less the cost of collecting it) or the value an advertiser would put on listeners. This is a common problem faced by sellers of advertising media. Wishing to obtain advertising revenue, they seek large audiences. One path to large audiences is through low prices to consumers, which of course lowers subscription revenue. A balancing is required. In the case of metropolitan daily newspapers, for example, about 20 percent of the revenue is from subscribers, 80 percent from advertisers. But many other publications have made different choices, all the way from free "shopper" newspapers to magazines with no advertising (such as *Consumer Reports*).

Packages for Sale, Bundles of Joy

One problem with charging consumers for programming has been that it is awkward to monitor the listening to or viewing of over-the-air broadcasts. Even cable television traditionally has sold the service as a package, with no attempt to monitor usage. Monitoring and charging are not, and never have been, impossible. But the cost of monitoring has remained high, on a per-transaction basis, relative to the prices that consumers are willing to pay for individual programs. Even though television viewers are generally willing to pay far more than advertisers for a given program delivered directly to them, it has been not profitable until recently to try to collect that money. Instead, television has been sold in packages, without usage-sensitive pricing. Of course, many services and commodities are bundled into packages and sold as units for reasons other than high transaction costs.

The problem of high transaction costs also plagues the Internet. Users might be willing to pay a small sum (say, a dime) to obtain an item of data or for the right to search a database. Many Internet information publishers offer service for free as a marketing experiment, but eventually they are going to want to charge. Unfortunately there is as yet no mechanism by which small sums can be collected efficiently and reliably from thousands or millions of users and distributed to information suppliers. What is happening instead is that suppliers are beginning to offer monthly subscriptions for wholesale access to their databases, again without usage-sensitive pricing. This creates a large impediment to commerce, however. There are many databases I am unwilling to subscribe to, but from which I am willing to pay for a single item now and then. For example, I am unwilling to pay $5 a month for real-time access to all prices on the New York Stock Exchange. I would gladly spend a dime each for weekly quotes on my ten stocks. Many databases have information of unknown value to most potential users; a monthly access subscription prevents efficient searching for information in these collections. Like the radio pioneers who decided how to package and market radio broadcasts, Internet content providers must decide whether to offer bundles of services on a subscription basis or to sell in small units, and in either case they must decide whether to include advertising.

Transaction and monitoring costs on the selling side are not the only reason that radio programs, like many Internet services, are offered as packages, sometimes called channels. For users or viewers to search out, select, and then arrange for delivery of preferred material is a costly process. Consumers value packaged material, like a daily newspaper, that reflects the services of an editor. Today the Internet offers many customized news delivery services. The user selects from certain categories of news, or news containing certain keywords. Tailored e-mail messages then deliver news items from those categories. These "push" Internet services offer an alternative to editorial services embodied in a daily newspaper or a daily television news broadcast. In neither case does the consumer sort through the thousands of news stories that are available each day.

Radio broadcasting started out using a single frequency. In other words, radio programs of necessity were originally "packaged" within a single channel through time. Had there been multiple channels, several competing radio broadcasters in each locality might each have started out with two or three channels of programming. NBC, for example, offered on one channel a schedule of radio programs designed to attract and keep certain large categories of listeners (listeners attractive to advertisers). Like program followed like program, so listeners need never leave their seats to change the station. Like the editor of a daily newspaper, NBC selected from among the hundreds of program ideas before it a menu designed to be attractive to its audience, or at least not so offensive as to bring the audience to its feet.

Anything You Can Do We Can Do Better

Another perspective on early radio development is provided by the concept of network effects (or network externalities). Network effects have several implications. First, a critical mass may be necessary before a service can begin to grow rapidly. It made no sense for investors to build radio stations until there were a number of radio sets in use, and it made no sense to buy sets until there were stations. A few experimental sets and stations existed for many years, until a point was reached when both could profitably expand. Such was the case with fax machines, and such may be the case with the Internet.

Second, obvious business advantages accrue from taking advantage of network externalities. Every time RCA sold a radio set, it increased the demand for NBC's programs and for NBC's audiences. Microsoft increases the demand for its operating system software (such as Windows) each time it produces a new application (such as Microsoft Word) that relies on that system, and vice versa. While it may be to the advantage of individual business enterprises to do otherwise, interconnection and compatibility give consumers and producers an opportunity to capitalize on network externalities without the burden of monopoly. It made business sense for RCA and GE to agree on standards and to make radios that could receive all stations, not just RCA stations or GE stations, respectively. Doing so increased the demand for sets and for stations.

Third, from the point of view of entrepreneurs venturing to offer a new good or service in an environment rife with network effects, it is essential that other producers make compatible or complementary products or adopt the new product or service as a standard. Other producers will do so only if they expect the new product to be successful, for compatibility is seldom achieved without costs. If technological change is rapid, it may be very difficult to predict success accurately. Therefore, the producer of a new good or service must use every means at its disposal to create the impression that the new product will be wildly successful, even before it is on the market. The result is a barrage of hyperbolic publicity aimed chiefly at other producers rather than at prospective customers. Rapid technological change generates hype all by itself, but the influence of network effects exaggerates the exaggerations. No doubt members of the industry recognize it for what it is. But the hype spills over to the nonspecialized press, to government officials and the general public, creating impressions and expectations that often are not realistic.

Economist Think

When discussing industrial organization, economists rely on such terms as markets, competition, market power, barriers to entry, and consumer welfare, each of which has a specialized meaning. The future of television and the Internet depends in part on factors best approached from the point of view of these industrial organization concepts.

For example, a market consists of all the products or services that are good substitutes for a given product or service, from the point of view of a given set of buyers. Sellers in such a market include not only those who presently sell the good or service but also those who would do so if a profitable opportunity arose. In the case of radio, there are three potential markets: a market for the sale of audiences to advertisers (or markets for the sale of particular types of audiences, say, males aged eighteen to forty-nine, to particular types of advertisers, say, auto manufacturers), a market for the sale of programming material to radio stations, and finally a market where stations compete for listeners by supplying attractive programs.

To illustrate the notion of counting sellers, consider the radio stations that employ a particular format, such as soft rock, in a given city. Is there a market for soft rock broadcasts, and if so who are the sellers? Soft rock may have good substitutes among other music formats for some listeners; other listeners may like nothing else. From the point of view of advertisers, the soft rock audience may or may not have desirable demographic characteristics. If soft rock is an especially profitable format relative to the formats of other stations, those other stations must ask whether they can change formats, and if so at what cost. It is not costly to change formats, and stations do it all the time. Therefore, the market for soft rock includes not only the current stations that play soft rock but also all the other stations that might switch to playing soft rock music.

Competition is the process by which each seller seeks economic advantage by better serving the needs of buyers, at the expense of its rivals. Each radio station seeks to deliver an audience bigger than that of its rivals, or one with a better demographic composition. Market power is a seller's ability to escape some or all of the discipline created by vigorous competition. If all the radio stations could get together and profitably fix advertising rates, they would achieve collective market power.

A barrier to entry is anything that keeps new sellers out of a market even though high profits are being made by the incumbent sellers. The necessity to obtain a license from the FCC is an excellent example of a barrier to entry, and this requirement has sustained the profitability of broadcasters for many decades, at the expense of consumer welfare. Consumer welfare is the aggregate satisfaction of buyers, generally measured by the difference between what they have to pay for goods and services and what they would be willing to pay. When markets are working well, competition produces greater consumer welfare than does either monopoly or government regulation.

Rents: The Commercial Pork Barrel

Economic rents are profits over and above what is necessary to keep a firm in business. Broadcast stations in the past earned economic rents attributable to artificial spectrum scarcity. Most of the rents, however, had to be turned over to the previous licensee. It is preferable, obviously, to keep rents. People pay lawyers, hire lobbyists, and make campaign contributions to that end. The problem with rents is twofold: they often arise at the expense of consumers, from some artificial or unproductive scarcity, and perfectly good resources are wasted fighting over them. Indeed, some economists view most of the regulatory activity that takes place in Washington (and in state capitals) as nothing more than the wasteful creation and distribution of rents—an activity that has no useful purpose other than defining employment opportunities for politicians and lobbyists.

The struggle to create, to retain, or to transfer rents is a major motivating force behind economic and political activity—even warfare. One reason radio succeeded was that the radio pioneers hoped to obtain rents. One reason the particular form of radio that we have succeeded was because the government protected the rents of the Sarnoffs and Paleys, providing incentives to invest in their form of radio. Rent seeking that takes the form of innovative and entrepreneurial behavior is not necessarily nonproductive; it may benefit consumers. Fighting over existing rents probably is entirely wasteful. A case in point is when multiple applicants battle it out through political contributions and legal fees to see which one will be awarded some monopoly franchise. Until about a decade ago, the government awarded and sometimes renewed all broadcast licenses in this fashion.

Living Dangerously

Finally, risk taking is at the heart of the development of new digital technologies. By their nature, the ultimate success of new technologies and business plans cannot be predicted with certainty. If they are to be developed, someone must risk time and resources by investing in them. Any sensible person or company must be paid to accept economic risk. Almost everyone prefers less risk to more, except when small sums are at stake. Most of us would prefer the certain receipt of

$1 million to the one-in-a-thousand chance to win a prize of $1 billion. Yet from an actuary's point of view these are equivalent propositions. The expected value (payoff times probability) of the two is the same.

In order to induce anyone to take a financial risk, that person has to expect compensation for taking the risk. The compensation is a higher expected profit. An investor leaves the safety of government bonds for the risks of the stock market only if she expects the return in the stock market to be greater than the return in government bonds. Every technology discussed in this book was, at the outset, extremely risky—much riskier than most mutual funds, for example. This was certainly true of radio in the 1920s.

Making Money: The Broadcasters' Business Plan

In the early days, radio broadcast stations developed a business plan, adopted later for television, that relied on four fundamental assumptions.

- There was a demand for broadcast entertainment that would cause people to purchase radios and to listen.
- There was demand by advertisers for access to listening audiences to hawk their wares.
- A supply of radios would be available from manufacturers at affordable prices.
- The expense of purchasing or producing sufficiently attractive entertainment programming would be less than the revenues from advertisers.

The business plan that resulted from these assumptions was straightforward: obtain or purchase broadcast licenses, build stations, purchase programming, convince sellers of consumer products that radio advertising is effective, perhaps with free trials, and wait for the radio sets to appear in the stores and the audiences to tune in. And it worked. At first programming was unsponsored or sponsored only by institutions such as orchestras. The expenses of commercial stations were paid by manufacturers of radio equipment: RCA, Westinghouse, and General Electric. By the late 1920s, commercial sponsorship in nearly its present form was prevalent. Until the 1960s radio and TV programs were produced and supplied to the networks by sponsors, who were responsible for their content. The present system of multiple sponsors arose in part from the quiz show scandals of the late 1950s and in part from the fact that advertisers learned to purchase portfolios of commercials in multiple programs, thus shifting much of the risk of low ratings to the network. Broadcast advertising became the most cost-effective way to reach mass audiences; the economies of scale were so great that the cost per listener was very low.

Radio pioneers did not feel it necessary to provide listening sets to their audiences. They relied on market forces for this essential element of their business. Nowadays it is a standard marketing ploy to offer free or reduced-rate consumer products to induce subscriptions to such services. That is why cell phones are offered at low prices or for free, together with a commitment to subscribe for some minimum period. That is why Netscape and Microsoft and other sellers of server software give away their Web browsers—they want to enhance demand for the servers that are browsed, so as to become a "standard" method of doing business. Of course, RCA, as the dominant force in the radio industry in the 1920s, made its profits from selling radios and therefore was motivated to supply free programming; Westinghouse and GE had the same motivation. But a number of other broadcasters had no stake in the equipment manufacturing business. Indeed, given the competition both in manufacturing and in broadcasting, it would not have been possible to continue indefinitely the promotion of equipment sales through the provision of programming. Single-channel radios tied to, say, RCA stations were not in demand. Hence programming efforts by one manufacturer had the effect of offering a "free ride" to other manufacturers.

Figure 4.2 illustrates the radio business plan. The radio station is in the business of selling audiences to advertisers. Audiences are attracted with programs. By affiliating with a network, the station shares the cost of programming with other stations, thus affording more popular programming. Also, by affiliating with a network, the station shares in sales to national advertisers.

There are interesting parallels with more recent events. Early videocassette recorders were expensive (well over $1,000) and were used chiefly to record television broadcasts. The business of selling or rent-

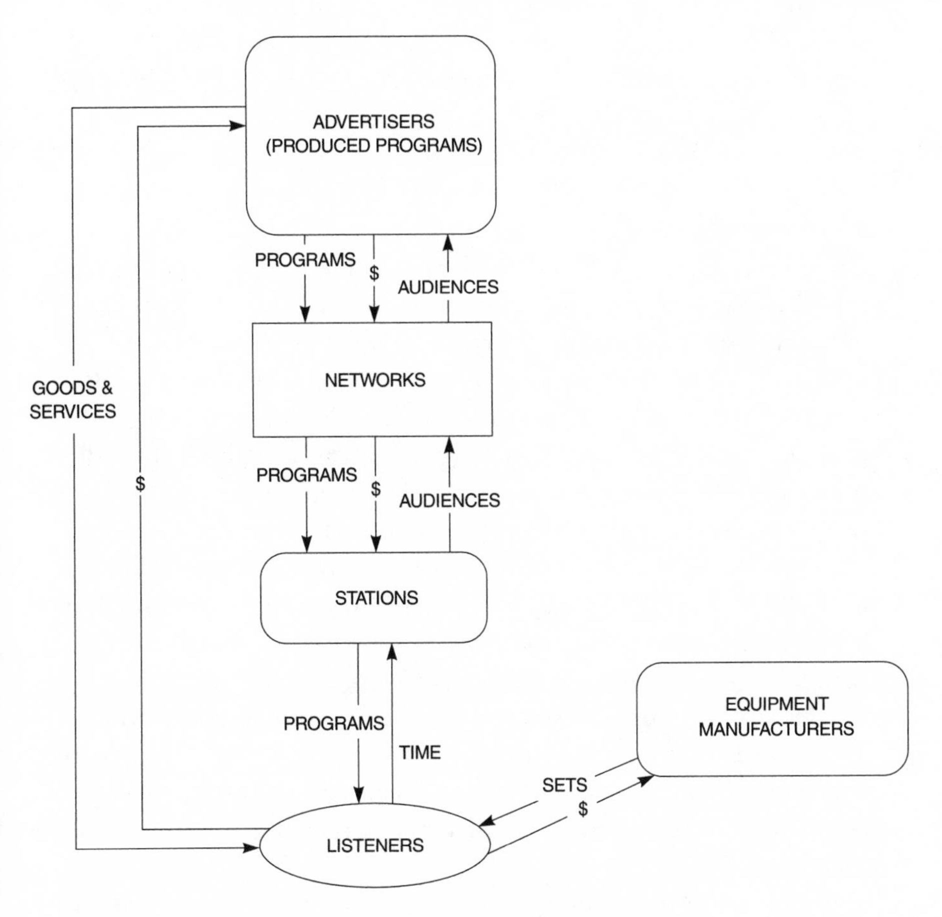

Figure 4.2 Organization of the U.S. radio industry, 1940s

ing prerecorded videotapes got started in part because entrepreneurs rented VCR equipment to consumers. As the volume of VCR production (and competition) increased, VCR prices fell, making a bigger market for rentals (and sales) of prerecorded cassettes.

For businesses that require the simultaneous development of complementary products, timing is everything, and coordination greatly reduces risk. Sarnoff coordinated the development of radio sets and radio broadcasting; the prerecorded videocassette industry expanded the demand for its services early on by exploiting the demand for pornographic videos, a need the Internet also serves, and for VCRs. Of course, to produce the whole range of complementary products required by such systems is very expensive and risky. That is one reason for all the "strategic alliances" announced each week in the computer and electronics industries.

The FCC Saves Sarnoff Again

At the time of World War II the Federal Communications Commission presided over a highly profitable amplitude modulation (AM) radio industry, with most of the structural features later associated with television. There were a handful of stations in each city, and a handful of national networks. Two technological innovations loomed. The most dramatic was television, a new medium backed heavily by radio broadcast interests, especially RCA. The other was an "improved" radio service based on frequency modulation (FM) transmission. Amplitude modulation impresses the data in a radio broadcast on the radio signal by rapidly varying the strength of the signal. An FM transmission accomplishes the same task by varying the pitch of the signal. For various technical reasons, FM broadcasts are less subject to interference than are AM signals, and their "signal to noise ratio" is higher. Therefore receivers of a given cost can produce better sound with an FM signal than with an AM signal broadcast at the same power level. Not surprisingly, FM is the dominant method of radio broadcasting.

In the case of FM radio stations, the FCC had an opportunity to expand the number of competing broadcasters and networks as well as to improve markedly the reception of radio signals. By 1945, there were 48 FM broadcast stations and about 500,000 FM receivers in the hands of the public. FM broadcasting faced the same chicken-and-egg problem that had earlier confronted AM broadcasters: it's hard to make money selling advertising until the public starts buying sets and listening, and it's hard to get the public to do that without attractive programming, which is expensive. Despite this difficulty, FM seemed on the verge of taking off in the postwar environment, a prospect that was discomfiting to established AM broadcasters. Worse, from David Sarnoff's point of view, the FM stations were broadcasting in a frequency band that he coveted for initial use by television stations.

To accommodate Sarnoff, the FCC decided to uproot FM radio, assigning it to an entirely different frequency band. This rendered all

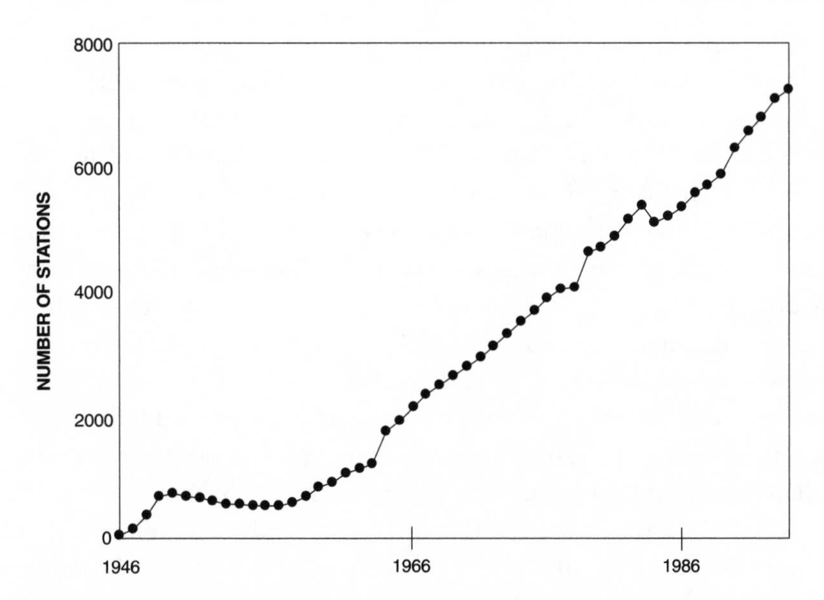

Figure 4.3 Growth of FM radio broadcast stations, 1946–1996 (TV & Cable Factbook, 1997)

the existing FM transmitters and receivers useless. The practical effect was to postpone for many years the commercial development of FM broadcasting (see Figure 4.3), and to protect the economic interests of AM broadcasters. The FCC offered an explanation of its action based on highly controversial technical considerations. It relied chiefly on the secret testimony of a Defense Department engineer, testimony that was later discredited. It is reasonable to infer that the decision to decimate FM was motivated largely by a desire to accommodate established broadcast interests such as those represented by Sarnoff.

Chain Radio and the Nature of Networks

Radio networks or "chains" were formed on an experimental and ad hoc basis in the early 1920s. The first full-time national radio network, NBC, was formed by RCA and its major stockholders in 1926. When RCA acquired AT&T's radio assets, the Radio Trust ended up with duplicate stations in New York. To avoid duplicate programming, NBC was split into two divisions, the "Red Network" and the "Blue Network." Each provided sponsored programming to radio stations across the country, in effect renting the time from their affiliates. As a matter of RCA policy, the Red Network carried the most popular programs and had the most listeners. In 1928 William Paley acquired the assets of United Independent Broadcasters and formed CBS, which for many years struggled behind NBC's two radio networks and did not equal NBC in revenues until 1948. Meanwhile, under pressure from the government, NBC divested the Blue Network, the weaker of its two networks, to a candy manufacturer, which formed the American Broadcasting Company (ABC) in 1943.

NBC's dual networks did not seek to serve the same audiences. To do so would have been foolish, for it would have reduced total advertising revenue. As a result, listeners enjoyed greater program diversity than they would when each network had a single channel.

One lesson to draw from this brief history of the development of the radio, and then the television, networks is the great power of incumbents in a regulated industry to shape and bend the adoption of new technology to their own ends. Programs are public goods. The genius of Paley and Sarnoff as entrepreneurs in the early days of radio was their recognition of the implications of that fact. Suppose a radio station in Philadelphia paid $100 for the right to broadcast the Philadelphia Symphony Orchestra and a station in Scranton paid $50 for the rights to a local musical group. The costs of broadcasting in Scranton could be reduced from $50 to zero simply by using the Philadelphia programming, which had already been paid for. The two stations could do this by forming a "chain," which later was called a network, linked together by AT&T telephone wires.

Networks at first chiefly had the characteristics of program-buying cooperatives, but as they covered larger areas of the country it became possible for them to sell "national" advertising. Subsequently, from the point of view of stations, networks performed the dual functions of program acquisition and national advertising sales. Of course, from the networks' perspective, the stations were merely local distribution outlets for programming and advertising, and the networks paid the stations "compensation" for the use of their time (see Figure 4.2).

As the radio networks grew larger and more powerful vis-à-vis their affiliated stations, disputes arose over the division of the profits from network broadcasting. The station owners' political influence outweighed that of the networks. Regulators consequently weighed in with restrictive regulations intended in theory to "protect" the stations from the economic power of the networks and in fact to give the stations a larger share of the profits. (In most cities the number of networks equaled the number of stations so there was no imbalance of power.) The regulations may have given the stations a larger share of the profit pie, but they had the effect of making the pie smaller, by outlawing efficient contracts between stations and networks. As part of this effort, the FCC enacted a "duopoly rule" that required NBC to divest one of its two radio networks. Today, of course, there are dozens of radio networks, many owned by the same entities.

Radio as Paradigm

For those eager to predict the future of television and the Internet, the story of radio holds many lessons. The problems that had to be overcome to bring about an organized and successful industry, to bend radio technology's many possibilities to a particular form, are fundamentally the same as those faced by the Internet as a potential mass medium. Some of the solutions to the early radio problems hold promise as solutions today. For example, the "strategic alliances" that dot the high-tech landscape are often intended to internalize network effects and to reduce risks. Many of the solutions require insight, investment, risk taking, and luck on the part of individual entrepreneurs. Others, however, require industry cooperation and government intervention.

Viewed in this light, Herbert Hoover's role in the development of commercial broadcasting appears ambiguous. Many of his policies appear today to be anticompetitive and anticonsumer in their intent and effects. At the same time, his policies undoubtedly made commercial radio a less risky investment and thereby spurred growth that, taken by itself, benefited consumers. The federal government today faces similar issues with respect to the Internet. A balance between promotion and free competition must be sought.

If the Internet industry decides to follow the path of the radio pioneers, it will form cooperative industry councils and seek government regulation of important aspects of its business, especially technical standards. One or a few major companies will provide, or appear to provide, important public services, such as service and equipment for the disadvantaged. In return they will be accorded a measure of protection from competition. The industry will grow more concentrated and the rate of innovation will slow. Of course, nothing compels industry leaders to take this path. Indeed, Silicon Valley has until recently been unsophisticated in dealing with Washington.

Conclusion

In the early 1920s the radio industry was in much the same chaotic state that digital media are today. No one knew which implementation of technology would prevail or what business model would succeed. Economies of scale, public goods, network effects, and other now-familiar economic forces bedeviled planning. The extent of consumer and advertiser demand was unknown. Grave financial risks faced the major companies. Entrepreneurs solved these problems in part by joining forces and by enlisting government regulators in their effort to limit competitive and technological risk. In addition, entirely new business strategies were invented.

CHAPTER FIVE

THE TRAGEDY OF BROADCAST REGULATION

If it were not for the failure of government policy, American consumers would have had much more and more diverse radio and television much sooner than they did, supported by vigorous competition. Sadly, not only has broadcast regulation been wrong-headed and harmful, it has been largely unnecessary.

Broadcast regulation began in earnest in 1927, when the Radio Act nationalized the airwaves. From the beginning, regulatory policies were heavily influenced by and therefore beneficial to the industry they regulated. This was by design. In close cooperation with Herbert Hoover and the Commerce Department from 1921 to 1929, the industry did much to encourage and to engineer its own regulation. Government intervention in those years may have reduced risk and encouraged investment in the new radio medium. For these reasons, some defend early government intervention. But continued regulation had no such good effects or rationale. Certainly the advent of television in the late 1940s was attended by none of the confusion, risk, and uncertainty that attended the birth of radio, and thus provided no justification for the restrictive measures imposed on the industry.

The government has almost always acted to restrict and restrain both competition and output in television markets in order to protect the economic interests of members of the industry. The government has permitted technological change to happen only very slowly for the same reason. In this respect the FCC has behaved no differently than various now-defunct federal regulatory agencies such as the Civil Aeronautics Board and the Interstate Commerce Commission.

Regulation is an invention and creature of the democratic process. Regulatory agencies never make important decisions in a political vacuum, as a court might. They consult with executive agencies, the White House, and several congressional committees (the House and Senate Commerce committees, among others) before announcing any major decision. Interested industry groups must lobby not only the agency concerned but also other agencies and elected representatives with a stake in the outcome as well. What comes out of this process is some sort of compromise, or perhaps consensus. And that, of course, is what representative government is all about.

The difficulty is that political outcomes, however procedurally democratic, are not market outcomes—and for economic affairs, market outcomes are generally better for consumers than political consensus. For that reason, ideally, the political process is substituted for the market process only when the market has failed badly, or would fail, but for the intervention.

As the FCC itself has demonstrated in recent years by gradually deregulating some parts of the television industry, there is no reason to suppose that unregulated television markets will fail. Regulators have probably known this for many years. But then, why regulate? The answer is that having begun to regulate, the government created a set of economic interests that stand to lose from deregulation. They stand to lose because deregulation means increased competition from new sellers and the destruction of incumbents' rents. So the rationale for continued regulation and the suppression of market forces is not market failure but the protection of clients whose existence or status is attributable to the government's prior actions.

Allocation of room in the airwaves for digital television stations provides a good illustration of these issues. Some policymakers have proposed giving every television broadcaster one free digital channel in exchange for its conventional channel. Then Senator Bob Dole, Representative Barney Frank, and other legislators opposed this so-called giveaway to broadcasters and argued that the broadcasters should be required to pay for their new digital channels. Is it a windfall to a broadcaster to be handed a free digital channel in exchange for its current analog one? There is no easy answer. On the one hand, the conventional channels were originally given to broadcasters without charge. On the other hand, present-day broadcasters are not the same entities who got those free licenses. Current broadcasters have had to pay previous owners tens or hundreds of millions of dollars for their channels. The original windfall was capitalized into the market value of each channel, and any owner after the first has paid full value for it. Thus, to take away the present channels while requiring broadcasters to pay for the new digital ones would be arguably an unfair "taking" of private property.

Digital television actually presents two problems: (1) allocating spectrum in an efficient manner that maximizes economic welfare and (2) ensuring that everyone involved is treated fairly. An effective solution to problem (1) would maximize the size of the pie available to be divided in solving problem (2). Often, however, there is no practical way to compensate those that lose from efficient changes. The lack of a mechanism to compensate losers is one of many reasons why the political process and regulatory agencies in particular are so resistant to change, even change that would make society as a whole better off. The issue arises in many other contexts. For example, an inability to decide how to compensate the owners of obsolete facilities has delayed the introduction of competition in the supply of telephone service and electricity.

Bigger Pie or Bigger Slice?

One of the easiest public policy choices is one that makes some people better off while leaving no one worse off. But seldom do such "Pareto superior" choices present themselves. Very often, the welfare of some can be advanced only at the expense of others. One way to approach the more difficult choices is to ask whether the pie as a whole is made bigger by some policy choice, or whether it is merely being divided in a different way. If the welfare of society as a whole is increased, it may be worthwhile to make the move, even though some groups are made worse off. Such policies are especially attractive if there exists a mechanism by which some of the winners' gains can be taxed away and used to compensate the losers. Too often, there is no such mechanism, and in a representative system of government it is not difficult for groups of potential losers to block action on plans that would hurt them.

For three-quarters of a century, the federal government has specified in great detail the way in which the airwaves can be used, for what purpose, and by whom. These rules fill 1,330 pages in the Code of Federal Regulations, and every one of them affects the ability of communication firms to compete and to adopt innovative methods of using the airwaves. And yet no change is possible without the elaborate and ponderous process of winning the government's approval.

Like other agencies, the FCC operates on two levels. One is formal, even ritualistic, following the dictates of due process as defined by the Constitution, the Administrative Procedures Act, and various court decisions. There are notices and comment periods, and reply comments, and petitions for reconsideration, and documents with titles like "Sixth Further Notice of Proposed Rulemaking and Notice of Inquiry in the Matter of the Assignment of Frequencies in the Fixed Satellite Service." Sometimes there are hearings, which may be organized either like court proceedings or like congressional committee hearings.

The other level is political. Congress created the Federal Communications Commission in 1934 and delegated to it certain regulatory functions (to regulate telephone and telegraph common carriers, allocate radio frequencies, license radio and television stations, and so on) that were too complex to handle through ordinary legislation. As such, the FCC, like other federal regulatory agencies, is an arm of Congress. The President appoints its members, who must be approved by the Senate, and Congress appropriates the agency's operating funds. The FCC's main political function is accommodating diverse interests. The building of coalitions of interests, usually blocking coalitions, is the meat and potatoes of regulatory policymaking.

Nowhere in either of these two levels of operation—the formal and the political—is there much room for sound economic and technical analysis of the policy choices that will best serve the public. This is not to say that we need agencies peopled by an elite cadre of technocrats or economists tasked to serve the public interest as best they can without regard to process or politics. That would be exceedingly undemocratic, and there could be no guarantee that the technocrats would get the answers right. Most of the time, the technocrats would have little or no basis on which to guess what the right answer might be. If regulation of the communication industry is required, the FCC is the kind of regulator we need and deserve.

But do we need to regulate the broadcast industry? What market failure is so serious in this industry (as compared, say, with book or newspaper publishing or computer software manufacturing) that the guiding hand of government is required to protect consumers against market outcomes? Very simply, there is none. Segments of the industry have substantial market power. But this power is protected most closely not by natural barriers to entry, but by federal and state governments, whose permission is required by new entrants. At present, the airwaves do not have well-defined private property rights that would prevent serious externalities in their private ownership and use. But the reason is not that such rights cannot be defined; it is that the government has prevented their development. Broadcast regulation is an emperor with no clothes.

Naked or not, the regulators are not about to run away and hide. The FCC will not go the way of abolished regulatory agencies like the CAB and the ICC. Any attempt to understand why the communication industry has the structure it has, much less any attempt to predict its future, must take the FCC and the political system whose interests it serves into account.

Uncle Sam as Mary Shelley

It is difficult to say much that is good about the role of regulation in the later history of broadcasting. So restrictive and distorting has been the effect of regulatory policy that it is almost impossible to imagine what broadcasting would have been like had it developed in a free market environment, except that there would have been a lot more of it a lot sooner. While it is possible to understand the underlying political motivation, the details of broadcast regulation in America are a Frankenstein's monster hardly more sensible than the Internal Revenue Code.

There have been two central difficulties. First, the government has assumed responsibility for determining all of the structure and much of the behavior of the broadcasting industry. This central planning role is not workable. Markets, even very imperfect ones, seem to do better than planners at serving consumers' needs, and markets are far more easily revolutionized by innovation than are regulatory fiefs. Second, because regulation is an element of representative government, its goal has rarely been efficient outcomes. Instead, the goal has been to reconcile and arbitrate interest group positions. But neither consumers nor those industries or technologies that would exist but for the rules can organize effective interest groups. Hence the government ends up catering chiefly to those economic interests that it has created or nurtured in the past. The result is a strong tendency to perpetuate the status quo, to protect economic rents, and to avoid undue competition.

The first of these problems is probably best illustrated by the collapse of Soviet economic central planning. The second problem can be illustrated by considering a few of the more famous broadcasting rules. Several have been mentioned already: the government's accommodation to RCA chief David Sarnoff in the matter of FM radio spectrum assignments after World War II, and the chain broadcasting rules set up to protect stations from networks. In effect, the chain broadcasting rules increased the cost and reduced the efficiency in the broadcasting industry. Although the rules shifted profits from networks to broadcasters, they also reduced the overall amount of profits to be divided. The network "duopoly" rule, mentioned in Chapter 4 in connection with NBC's divestiture of the Blue Network, probably had an adverse effect on industry performance as well. When one company owns two networks or channels, it tries not to duplicate the programming on them or to serve the same audience. Instead, as with the BBC, each channel will be aimed at a different kind of listener or viewer. For this reason, the network duopoly rule probably reduced program content diversity in broadcasting, although it may have increased competition in advertising markets.

The chain broadcasting rules were upheld on appeal to the Supreme Court in the 1943 Networks case (National Broadcasting Company, Inc. v. United States, 319 U.S. 190 (1943)). On this occasion Felix Frankfurter invented out of whole cloth the scarcity rationale for broadcast regulation, reading into history the false assumption that government regulation (beyond the Radio Act of 1912) was necessary to prevent "chaos" on the airwaves. Behind Frankfurter's synthetic justification for broadcast regulation, and behind the views of many influential lawmakers and opinion leaders then and since, lies a deep suspicion that broadcasting is too powerful a social force to leave unregulated, whatever the economics of it may be. Frankfurter wrote his opinion at a time when the public well remembered the power that radio conferred on demagogues such as Father Coughlin and Huey Long. Other nations have generally taken a similar view; indeed, until recently, most kept broadcasting a state monopoly.

The notion that broadcasting has too much social importance to be left to the market is ironic: the fear of powerful media arises from the fact that the broadcast media have been so highly concentrated—a consequence of regulation, not of any natural feature of radio or television technology. Spectrum could have been bought and sold with no greater difficulty than land, and is no more scarce. Now that concentration in broadcasting has been greatly reduced, probably the greatest remaining "demand" for regulation comes from the industry itself, as it seeks to manipulate the government to acquire or retain economic rents.

The government has a variety of "ownership" rules, some of them now codified by the Telecommunications Act of 1996. There are limitations on the percentage of subscribers any cable operator can serve. There are limitations on the percentage of the population that jointly owned TV and radio stations may serve, both nationally and in any given city. Some of these rules make sense as antitrust policy, although the same result could probably be reached through direct application of the antimerger law (Clayton Act §7). But many make no sense at all, reflect the obsolescent concern with media social power, and simply inhibit industry performance. For example, there is no anticompetitive effect that arises from ownership of broadcast stations in more than one city, and probably there are economies in such ownership. There is no reason for this limitation. The same applies with particular force to restrictions on the sizes of cable MSOs, a subject discussed in Chapter 7.

Two particularly painful episodes of broadcast regulation began in 1970. The FCC adopted the Prime Time Access Rule (PTAR) and the Financial Interest and Syndication Rule (FISR). Just as the chain broadcasting rules were at bottom an attempt to shift profits from networks to stations, these rules were attempts to shift profits from networks to Hollywood studios and other programming interests by eliminating more efficient competitors.

The Prime Time Access Rule made it unlawful for any network-affiliated TV station to show network programs (as opposed to programs purchased from syndicators or produced locally) between 7:00 P.M. and 8:00 P.M., except for a half-hour of network news. This bizarre regulation spawned a programming industry that was devoted to cheap game shows and made stars of Vanna White and Alex Trebeck. The network programming that was replaced was of higher "quality" in the sense that more viewers watched it. (The reason that network affiliates previously broadcast network programming in this period was that network programming generated greater profits.) PTAR reduced overall TV audience size.

The Financial Interest and Syndication Rule made it unlawful for any network to have ownership interests in the prime-time series programs that it aired or to engage in the business of syndicating network reruns to TV stations. This rule made it impossible for the networks to finance or underwrite program production in exchange for an equity interest in the show. The Hollywood studios stepped in to provide such financing, but at a higher cost than the networks. (The studios individually were smaller than the networks and thus less efficient risk bearers. In addition, the networks were in a position to optimize program schedules, which the studios were not.) The effect was to increase the cost of producing programs, while increasing the profits of Hollywood studios.

Both PTAR and FISR were repealed in 1995, the latter after a scathing analysis by the Seventh Circuit Court of Appeal. The government's own analysis of the rules at the time of their repeal admits their adverse effects. Starting as early as 1980, staff studies had repeatedly condemned these rules, and they were universally treated with derision by scholars. And yet it took twenty-five years to repeal rules that were supported by no one but the Motion Picture Association of America (MPAA) and related Hollywood interests, including a certain former president of the Screen Actors' Guild named Ronald Reagan. The point is not to disparage the MPAA, which was after all remarkably effective; it is to highlight the process that permits small groups with strong interests to govern regulatory outcomes, especially when it comes to preventing change.

The Internet: A Fat New Target

On July 1, 1997, President Bill Clinton announced a new government policy toward the Internet. The policy was characterized as "hands off"—that is, minimizing government regulation and taxation. At the same time, the policy was promotional. The White House championed the Internet as an engine for the growth of transactional commerce as well as a vehicle for U.S. export sales of related computer equipment and software. The Internet, the President implied, would bring extraordinary high-tech benefits to American citizens—praise reminiscent of the pronouncements made at the time cable television was "deregulated" in 1984.

In spite of the government's "hands-off" pronouncement, regulation surely looms in the future of the Internet, for several reasons. First, the development of video or other entertainment delivery on the Internet will economically threaten the current television industry, which will quickly call upon the government to bring the Internet further within the regulatory ambit. The effect of this would be to attenuate the Internet's ability to threaten broadcasters—by no means a far-fetched possibility, as the cable industry knows to its chagrin. Broadcasters were able to delay the growth of cable for years by such means.

Second, the government may be the solution to some of the problems of excessive risk and lack of standardization that today delays the commercial exploitation of digital technology. In other words, the government may do for the Internet what Secretary Hoover did for radio—make the technology safe for business. Just as they did in the 1920s, major industry players may seek government help.

Third, boisterous hyperbole is an important element of successful new Internet technologies. The government's overblown prediction of what the Internet can achieve, however, has created expectations that the Internet industry may be unable to meet. The government is prematurely licking its chops at the prospect of the social benefits, including tax revenues, that will result from Internet growth and commercialization. Some of those benefits are already being distributed to schools, hospitals, and the poor by overeager politicians. Businesses associated with the Internet are benefiting from a variety of preferences mandated by the government in order to encourage the growth of digital media. For example, Internet service providers and users are spared telephone access fees that might otherwise be imposed on them. If expectations are disappointed (and perhaps even if they are not), the government is likely to become disenchanted or impatient and to demand something in return, especially if there is an identifiable commercial enterprise or group to blame for the lack of performance.

The first forays into Internet regulation have already occurred: the 1996 Telecommunications Act mandates subsidies for Internet access by certain groups at the expense of certain other groups, the national security agencies have attempted to impose "key escrow" requirements on users of cryptographic software, and (in September 1997) the government proposed requiring that personal computers be equipped with the "V-chip," a move intended to regulate objectionable TV programming.

Conclusion

Restrictions of competition and limitations on spectrum availability may have facilitated the early development of radio. But these government policies did not cease after the developmental period, and they have had significant adverse effects on consumers for more than half a century. First radio and then television channels were limited, leaving unsatisfied consumer demand for programming. The legal establishment succeeded in inventing the required rationalizations. Such anticonsumer policies have not arisen simply from errors of judgment; they are a reflection of the workings of the political process, which requires regulators to take account of important economic interest groups. New digital media such as the Internet are not immune from this process.

CHAPTER SIX

THE EVOLUTION OF BROADCAST TELEVISION

The development of commercial radio in the United States in the two decades after 1920 served as a paradigm for the development of television. The broadcast technologies and production processes were similar. The form and basis of government regulation were identical. And the business plan that led to commercial success in television was borrowed wholesale from the radio experience. Indeed, many of the commercial pioneers who developed television—notably RCA's head, David Sarnoff, and CBS founder William Paley—were the same people who had developed radio.

The form of broadcast television that developed in the United States, like the form of commercial radio, was an advertiser-supported entertainment medium. It relied chiefly on sources of programming that were national but on transmitting stations that were local. Each viewer, therefore, had only a handful of options.

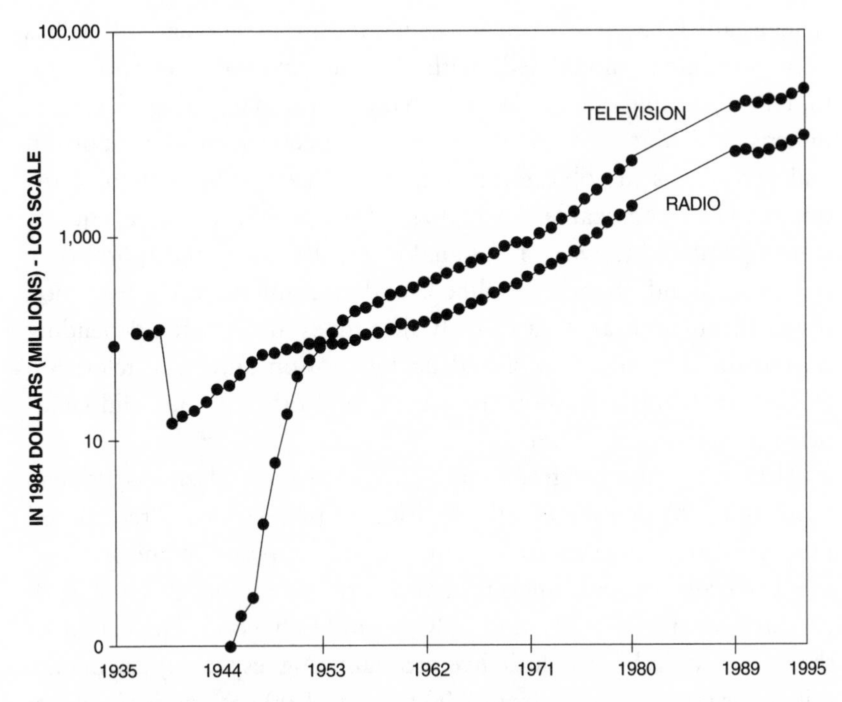

Figure 6.1 Broadcasting industry revenues, 1935–1995 (U.S. Department of Commerce, various years; data not available for 1981–1988)

The electronic recording and transmission of video images was invented by a Russian, V. K. Zworykin, in 1923, and independently by an American, Philo Farnsworth, in 1927. Television first came into commercial use in the United States as a mass medium after World War II. It quickly became commercially important, outstripping radio within a few years of its introduction (see Figure 6.1). In the late 1940s and early 1950s, a television set was a prestige consumer good, and television viewing was as novel and exciting an activity as "Web surfing" is for many people today.

The Basics

Television today works by means of a glass lens that focuses images on a light-sensitive semiconductor. The image is scanned at 480 active lines per frame and 30 frames per second. (Actually, each frame is composed of two interlaced parts, with alternating lines, transmitted sequentially.) The video camera generates an analog signal that has been imprinted (modulated) with this information. (A representational oil painting of an object is a two-dimensional analog of a three-dimensional object, varying in certain respects, such as proportion and texture, as the object varies, and thus capturing enough of the object to be recognizable. An analog TV signal is the electrical equivalent of paint and canvas.) The signal is translated into the appropriate frequency band, greatly amplified, and transmitted via a television tower throughout an area roughly fifty miles in diameter, depending on terrain. The signal is picked up by rooftop antennas, television rabbit ears, cable television companies, and other means, and channeled to millions of TV sets.

Although some programming, such as news and sports, is live, more than 90 percent of all television is prerecorded. Prerecorded programs are delivered to stations on videotape or in the form of satellite transmissions. Such programs originate either in New York (in the case of ABC, CBS, and NBC) or in Hollywood. The networks themselves do relatively little live programming, especially during the critical prime-time hours from 7:00 P.M. to 11:00 P.M.; network entertainment programs are purchased from Hollywood producers for prices that have sometimes reached $14 million per episode.

Making Money

Conventional broadcast stations earn profits by selling audiences to advertisers. The size and to a lesser extent the demographic composition of the audience are what count: more eyeballs, more revenue. To generate audiences, of course, broadcasters must offer inducements to viewers in the form of programming.

Figure 6.2 is a greatly simplified diagram of the organization of the television broadcast industry. "Advertisers" include local, national, and network advertisers, and "program suppliers" include syndicators, sports leagues, and Hollywood studios. The arrow showing "programs" going from stations to viewers stands for all the many ways in which that can happen. The FCC and other federal agencies have been omitted, although they play a crucial role. The diagram differs from the diagram of the organization of the radio industry in the 1940s (Figure 4.2) in two important ways: the decreased role of

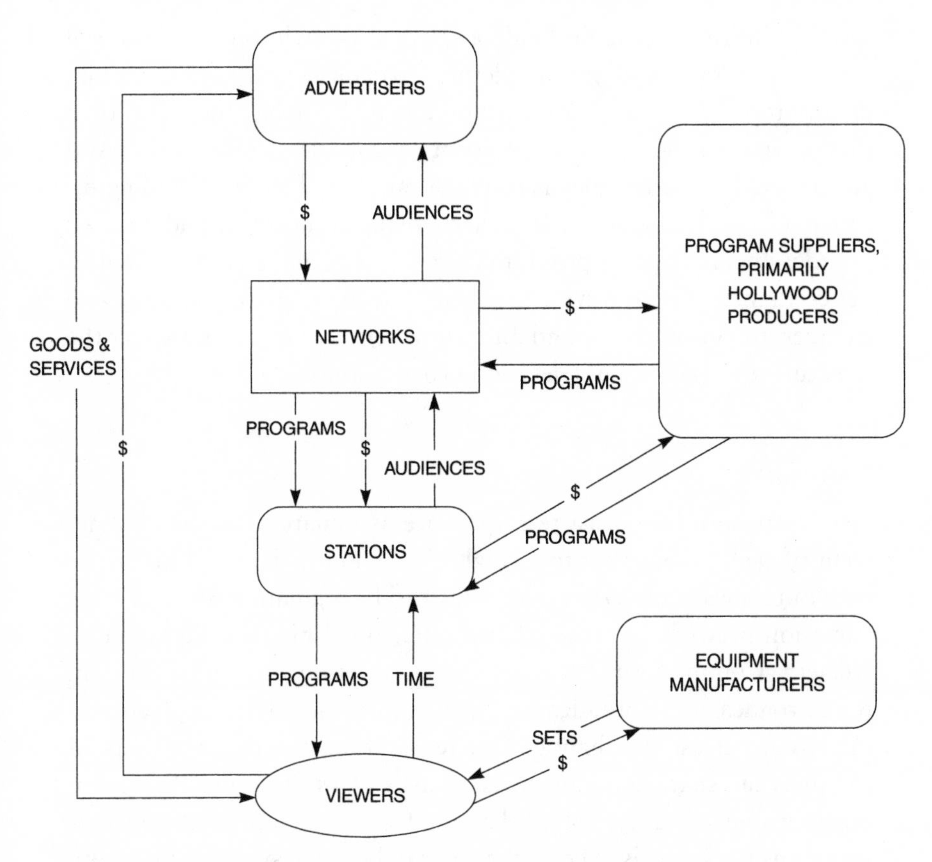

Figure 6.2 Organization of the television broadcast industry

advertisers (who no longer supply programming) and the very significant role of Hollywood. Radio (and early television) advertisers were responsible for producing much of the programming that was aired. They bought time in units of hours, not seconds as now. The locus of radio program production was New York. By the late 1950s, TV entertainment programming was being produced on film in Hollywood and sold directly to the networks.

The broadcast television industry stays in business only by delivering a cost-effective product to advertisers and viewers. On all fronts it faces competition. Programmers have alternatives to television. Instead of selling their programming to TV stations or networks, they can make motion pictures and obtain money directly from consumers through theater tickets and videocassettes; in addition, they can sell original programs and motion pictures to cable networks and satellite distributors. Viewers also have alternatives to television. The time they devote to viewing can be spent in countless other ways. And advertisers have other means to reach target audiences, including direct mail and newspapers. In order to stay in business and succeed broadcasters must offer programmers, viewers, and advertisers a deal better than the alternatives. Over time, technology, regulation, and changes in consumer demand alter these terms of trade and affect the structure and viability of television broadcasting.

Packaging

Few features of broadcast television are as unnatural as the confinement of each broadcaster to a single channel in each area. Experience with outcomes in markets not constrained by regulation, such as cable television networks, teaches that multichannel broadcasting is more efficient. Common sense teaches the same lesson. It is more efficient for a broadcaster to broadcast a "package" of channels in a given area than to broadcast one channel. One reason for this is that a package of channels can share common costs, such as news gathering. Another is that consumers can select packages of channels rather than single programs or channels. The packagers can be cable systems, for example, or satellite systems such as DSS. These packagers compete in selecting popular portfolios of networks that will attract and retain subscribers. Packagers perform an editorial function not unlike that performed by newspapers, which select and print each day a small fraction of all the news that is available. Readers benefit by, in effect, hiring an editor to select an interesting and attractive package of stories from the impossibly large number available.

Cable television systems and satellite systems offer very large packages of channels to capture the economies of editing and packaging. They sell only the most expensive movie channels and sports packages à la carte. Presumably, if viewers were offered a chance to buy monthly subscriptions to individual offerings such as the Cooking Channel, less money would come in than if they offered viewers a package of dozens of individual program channels. One reason for this is that people may be unwilling to pay to subscribe to a channel for which they have only a contingent demand for occasional programming. C-SPAN, which covers congressional debates and hearings, may be an example. Viewers might be willing to pay for a particular program when the occasion arose, but not (much) for the right to get all such programs without further charge. Even when your favorite TV series is scheduled to be aired, it may turn out that a particular episode is one you have seen before or don't like; only by sampling the product can you appreciate its value, and this reduces the demand for packages of such programs that must be purchased in advance. Systems like the pay-per-view movie services found in many hotels, which allow a five- or ten-minute free sampling period, might be more efficient than a subscription, if technical and transaction costs could be overcome.

Offering packages of programs or channels for a single price does have two attractive features. First, in some circumstances it may permit the supplier to extract a larger fraction of the consumer surplus or willingness to pay of users, and therefore (in a competitive market) to increase output. Second, many consumers place a value on not having to worry about usage; these consumers would be willing to pay something to avoid having a per-unit charge for their consumption of certain services. This, together with transaction costs, may explain the popularity of flat-rate pricing of local phone service. America Online (AOL) in 1996 apparently believed that it would gain new subscribers by eliminating usage charges and changing to a flat monthly rate independent of usage. AOL was correct, but the decision nearly scuttled the company: usage by new as well as existing subscribers immediately far exceeded available capacity, and quality of service deteriorated accordingly until additional capacity could be built.

Packages of programs, or of channels, also offer producers a chance to avoid the duplication that would take place if each channel or program were offered by a separate competing producer. It used to be said of the three major broadcast networks that they were all alike, duplicating one another's programming. This was one dimension of FCC chairman Newton Minow's famous 1961 complaint that television was a "vast wasteland." If three independent channels are vying for the largest audience to sell to advertisers, they will each tend to offer programs that appeal to the same largest group of viewers. If each broadcaster controlled several channels, however, it presumably would not duplicate programming within its own bundle of channels. Multichannel broadcasting would lead to greater diversity.

If programmers will avoid duplication of program types within the bundle of channels that each controls, what explains Ted Turner's Cable News Network (CNN) service, which was bought by Time Warner (along with the rest of Turner Broadcasting) in 1995? CNN offers four channels with a great deal of duplication of content among them: CNN, CNN-Headline News (CNN-HN), CNN-Financial News (CNN-FN), and CNN-Sports Illustrated (CNN-SI). The answer lies in the advantages to be gained from maximum exploitation of the fixed costs of producing programming, in this case news. After a news story is gathered for CNN, it costs almost nothing more to air it on CNN-HN, and if appropriate, on the financial or sports channels. In fact, failure to exploit such economies fully in this way would give CNN higher costs relative to ad rates than those of its competitors.

Virtually every product or service is really a package or assembly of component products and services whose association is dictated by cost or demand considerations, or both. An automobile has thousands of parts that can be purchased as a unit (a car) or individually, as spare parts. Packaging in the television business is no less natural or more remarkable. Traditional regulation has in many cases prevented the formation of efficient packages of video products, however.

The Federal Plan for Television: Frankenstein Returns

The government initially determined in 1946, on the basis of its own forecasts of demand and the industry's, that the equivalent of thirteen channels (6 megahertz each, but not all contiguous) should be allocated for television in the United States as a whole. It later reduced the number to twelve channels on the VHF (very-high-frequency) band. Because several of these channels were adjacent to one another in the spectrum, the practical maximum number of channels that could be used in any given area was seven. (All twelve channels cannot be used in a given area because adjacent channels interfere with one another. Regulators originally authorized seven VHF stations, the maximum, in New York and Los Angeles.)

Television signals in the VHF band travel in a straight line ("line of sight") from the transmitter to the horizon as viewed from the transmitter. On flat terrain, with conventional antennas and broadcast power, that distance is about 50 miles. Hence, a given frequency can be reused if the distant signal does not overlap the first, meaning that transmitters on the same frequency must be at least 100 miles apart. (Such frequency reuse is an important means of conserving spectrum, and it is a feature of many wireless communication services—for example, cell phones.)

When the government authorized seven VHF television channels in New York City, no other city within about 100 miles could have its own local station on those channels. The government could have authorized seven channels in a large number of such major regions of the country. Alternatively, the government could have attempted to maximize the number of communities that had at least one local station, an effort that would have minimized the number of television channels that could be received by any given consumer. This could be justified if it was thought that viewers placed a much higher value on local programs than on regional or national ones.

Instead the government took a middle course, arranging assignments so that most television viewers could receive three signals. The decision to make so few choices available to most viewers was justified by emphasizing the value of localism in programming. In fact, it turns out that viewers have very little interest in local entertainment programming, which is hard-pressed to compete with national programming for audiences. (There is, however, a stronger demand for local news and sports.)

Awarding licenses to television broadcasters was an intensely political process. It was generally accepted that television broadcasting would be profitable, and that the licenses would turn out to be at least as valuable as the AM radio licenses issued in the twenties and thirties. Radio broadcasters were the obvious choice to operate the new television service, because of their technical expertise. Further, the radio broadcast networks, especially RCA/NBC, had invested heavily in promoting the new technology. The FCC's decision to allocate the airwaves so that most viewers would receive three channels meant that most individual members of Congress had at least one license in their district, and thus an opportunity to participate in the assignment process. The government awarded licenses free of charge to those who applied; where competing applications were submitted, license awards were based on the "comparative merit" of the applicants. Comparative merit gave heavy weight to existing radio broadcasters, newspaper publishers, and those with strong congressional support. The monopoly television station license in Austin, Texas, awarded to Senator Lyndon Johnson's wife Lady Bird is a famous example.

There was no technical reason why the government had to limit the number of television channels to twelve, or the maximum in each city to seven. The FCC claimed it did so because of two concerns. First, it did not want to reserve a portion of the airwaves that would lie fallow for lack of demand. Second, it did not want to authorize so many new television stations that there would be excessive competition for advertising dollars and insufficient attention to public interest programming. In the end, however, one does not go far wrong in assuming the commission essentially did what Sarnoff proposed. What Sarnoff proposed was to maximize the economic rents that would be enjoyed by incumbent (radio) broadcast stations and networks as they moved into the new television medium.

It quickly became apparent that television would be a great success and that much too small a portion of the airwaves had been allocated for the new service. In 1953 the FCC responded by allocating seventy new television channels in the so-called UHF (ultra-high-frequency) band. Both in 1953 and later, the commission recognized that the new UHF broadcasters would have great difficulty competing with the established VHF broadcasters. The VHF signals were technically superior, and in any case few TV sets were equipped to receive UHF signals.

One solution to the "handicap" facing UHF broadcasters was "deintermixture," making the television stations in any given city all UHF or all VHF, which would eliminate any signal-quality distinction and ensure that all television sets would be equipped to receive UHF as well as VHF signals. Deintermixture also would have meant a substantial increase (at first, probably a doubling, from three to six) in the number of competing TV broadcast stations and viewer options in each community. Naturally, existing VHF broadcasters opposed such a scheme, and they succeeded in turning aside the deintermixture threat.

Instead, to solve the UHF problem the FCC asked Congress for the authority to regulate the manufacture of TV sets. Based on that authority, the All-Channel Receiver Act, the commission in the 1960s required that all televisions be equipped with UHF tuners. Later, in the 1970s, the commission required that UHF tuning dials have "detents" or click-stops at each of the seventy channels. (The continuous, radio-like UHF tuning knobs were thought to put off viewers.) None of this was very effective in promoting the development of UHF television, and it was not until cable television, electronic tuners, and remote controls came into wide use in the 1980s that the UHF handicap began to disappear. (Curiously, however, the 1996 Telecommunications Act limits television station ownership to 35 percent coverage of all U.S. households—unless the stations are UHF, in which case 70 percent is permitted.)

The UHF television experience invites comparison with current developments in digital television. All of today's television broadcast media (including multichannel multipoint distribution systems and direct broadcast satellites as well as any modern cable service) except VHF and UHF require the same sort of set-top box as did early UHF broadcasts. Future transmission media such as LMDS will also require such boxes. But set-top boxes are often the single most expensive component in a television transmission system because, unlike the transmission paths, they cannot be shared by multiple users. One way to deal with this problem is to induce TV set manufacturers to build some or all of the necessary electronics into the sets. Television manufacturers would be happy to do this if doing so would enhance the demand for sets in a way that more than offset the costs. But demand will not be significant for new media just entering the market. What cannot be achieved in the market can often be achieved at the FCC. It would not be surprising to find the FCC considering imposing TV set standards designed to ease the introduction of new transmission technologies. This is almost certain to happen with digital television, regarded as the commission's own brainchild.

Television Broadcast Networks: Radio Offspring

Television networks and radio networks operated in exactly the same manner, subject to the same rules. The CBS, NBC, and ABC radio networks spawned the CBS, NBC, and ABC television networks. Dumont, an early but weak fourth network, failed, and ABC itself struggled for many years. The proliferation of broadcast networks we have today is a very recent phenomenon. Fox began broadcasting one evening a week in 1987, and Paramount and Warner Bros. each started a broadcast network in 1995.

What makes a strong network or a weak one, and what determines how many networks can survive? The answer has to do with the public-good nature of TV programming. Consider two networks in competition. One has affiliates reaching 99 percent of all television households. The other has affiliates reaching 90 percent. Suppose each spends the same on programming and therefore is equally attractive to those viewers who can receive its signals. The first network will be able to make profits far higher than the second, and it may even be able to price its advertising in such a way as to operate at a profit while forcing the other network to operate at a loss. There are two reasons for this, aside from the obvious fact that the first network has more viewers. The first network has costs per viewer 10 percent lower than does its rival. And some advertisers, because they prefer to reach as many potential purchasers as possible without duplication, may be willing to pay more per viewer for the larger audience. The second network will be at a disadvantage that is difficult to overcome.

Therefore, commercial success for a broadcast network depends on having access to the same potential audience as its rivals. (The same is true of any industry distributing public goods. Drug companies, which spend vast sums on research for new products that cost very little per pill to manufacture, earn far higher profits if they can sell to world markets than if they are confined to their home countries. Software publishers face the same problem, which explains the recent resistance of U.S. computer companies to the government restrictions on export of certain encryption software.) Absent such access, the network must seek a niche in which it is not competing directly with advantaged rivals. This has been the strategy, for example, of the Fox network, which has specialized successfully in untried and inexpensive program types and younger audiences.

For many years, ABC was at a disadvantage next to CBS and NBC because its affiliated stations reached only about 92 percent of the country, whereas CBS and NBC reached 99 percent. Further, many of the ABC affiliates were UHF stations that viewers could receive only with difficulty, a problem directly attributable to regulation. The government restricted the number of broadcast licenses, reducing the potential for more stations and more networks. Later, in the 1980s, as the FCC began to authorize more VHF stations and as the UHF disadvantage declined, more networks appeared.

Regulatory restrictions limited the number of broadcast networks, but government policy is not the only determinant of the number of networks or the quantity and quality of TV programming. Government policies aside, there is as much television as there is advertiser demand for audiences and audience time-demand for programs. Over the years aggregate advertising expenditure per capita has increased as the economy has grown, and of course population has increased. Both factors have promoted growth in the number of networks.

Today new factors are coming into play in determining the number of television networks that distribute primarily through broadcast stations. One is the use of viewer dollars to support programming. Another is the appearance of rival distribution systems, some with the potential to do what TV networks do, but better.

Backward Broadcasting

Conventional television undoubtedly is the most backward of all electronics industries. The technical standards governing broadcasting are set by the FCC and can be changed only after formal rulemaking proceedings that are uncertain and lengthy. For example, standards adopted in late 1996 for digital TV broadcasting followed a decade-long consultative process. Because of rulemaking constraints, there have been few major changes—improvements that enabled fundamental alteration in the structure of the industry—in the technology of television broadcasting since the 1950s. Digital television, if it catches on, will be one. Other landmark changes include color television, the integration of UHF tuners, detent tuning, remote controls, and the use of the vertical blanking interval.

Color television did not become possible until the FCC adopted in 1953 a standard sponsored by CBS. (Earlier, it had settled briefly on a rival, RCA-sponsored standard that proved unworkable.) The commission's quixotic policy to promote UHF television hindered the speedy development of new viewer choices. It is impossible to know what broadcasting innovations might have taken place if manufacturers and broadcasters had been free to experiment with new products and services.

Revolutionary Remotes

Remote controls had a revolutionary impact on network programming strategy. Remotes at first, in the 1960s, were luxury items that did nothing more than heavily and slowly rotate the old mechanical tuners of the day. But by the early 1980s a large proportion of new television sets were shipping with electronic tuners and random access remotes. Channel surfing had arrived (see Figure 6.3).

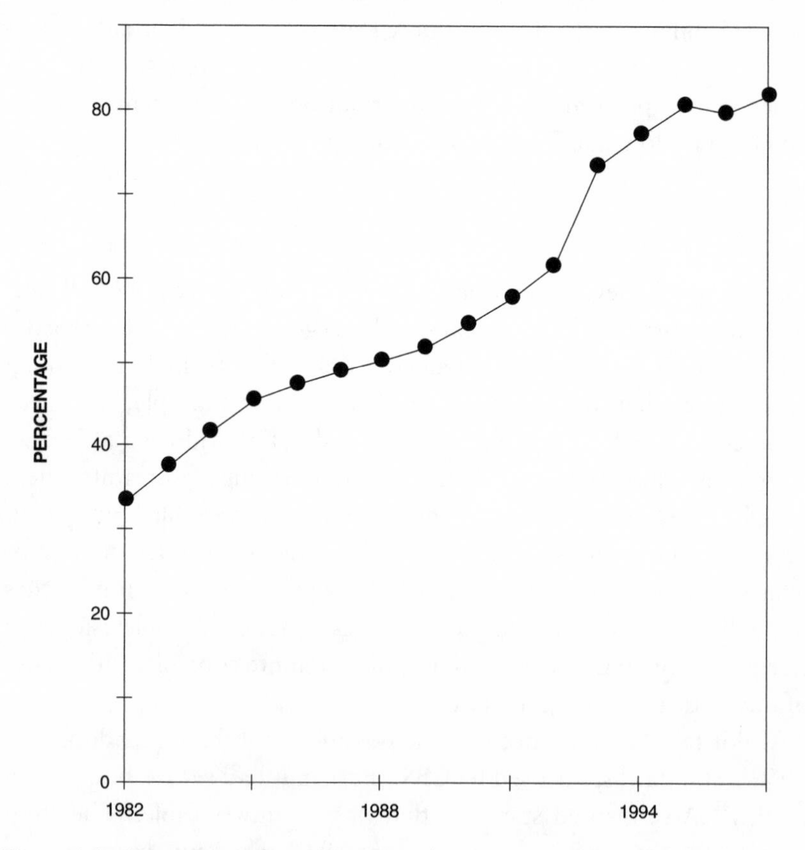

Figure 6.3 Percentage of new TV sets with remote controls, 1982–1997 (Gigi Information Services, 1997)

The freedom to channel surf gave viewers the opportunity to avoid oppressive commercials. More important, it ended or greatly reduced the effectiveness of the "audience flow" strategies of network programming, which relied on the passivity of viewers. (Programmers figured viewers would stay in their seats unless provoked—a captive audience as one program on the channel flowed into the next.) Now, with channel surfing, both commercials and programs have to stand on their own merits, without as much assistance from what has preceded. The random access TV remote control, which can move from one channel to another without passing through those in between, presages the Web browser, which can banish the boring known in favor of the potentially exciting unknown with the click of a mouse button. Or will it? For many viewers television is a passive activity, valued in part because it is bereft of thought or the need for frequent considered choice. Web browsing is the opposite. Maybe TV channel surfing and Web browsing are different activities that satisfy different needs.

VBI: A Drop in the Bucket

When a TV set flashes one frame of a TV picture on the screen, an internal electron gun traverses 480 lines in 1/30th of a second. When the beam reaches the bottom of the screen, it travels back to the top to begin the next frame. (As noted above, each frame is sent in two parts, with alternating or "interlaced" lines.) Although 480 lines appear, more than 500 lines are sent. There are 42 lines per frame that do not show up in the picture, although one can see their representation as the black bar between frames that appears when the picture loses vertical hold and "rolls." Some of these lines are required for timing and alignment of the video signal, but 24 of them are not used for any video purpose. The overall 42-line interval is called the "vertical blanking interval" (VBI). It has a usable information-carrying capacity, according to some experts, of up to 150 kilobits a second per channel, faster than the fastest analog modem and faster than today's digital (ISDN) telephone lines, but one-way, of course (the usable portion of the VBI is shown in Figure 6.4).

The VBI is like an empty millimeter at the top of a bucket of

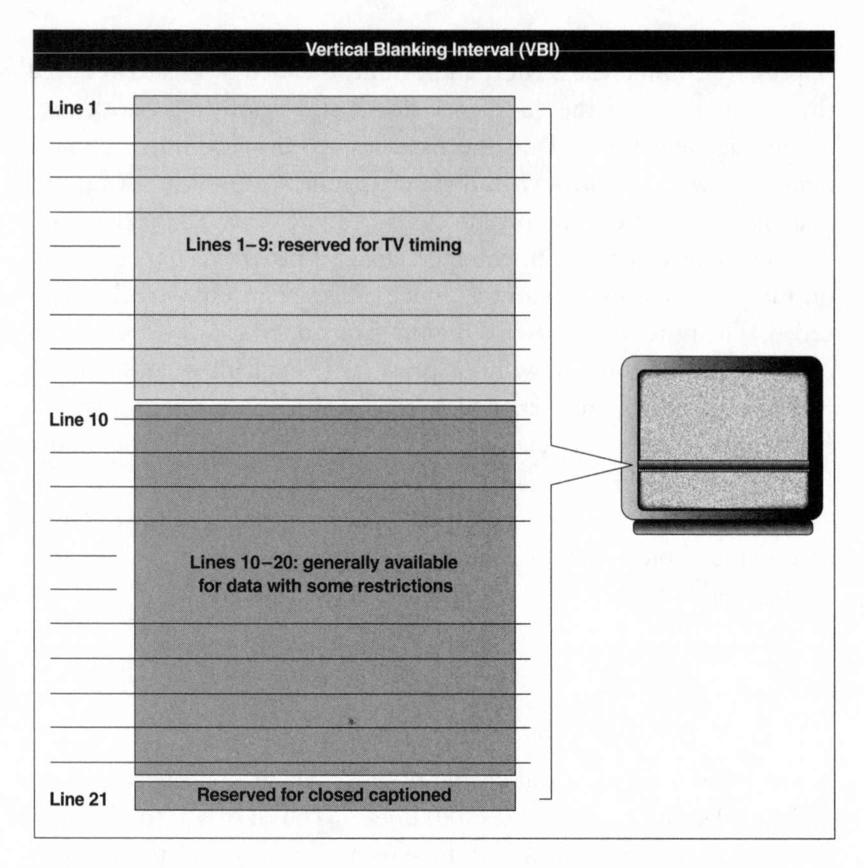

Figure 6.4 Vertical blanking interval (Intel, 1998; used by permission)

water—there is room for more water (data) in the bucket. Television stations use the VBI for a variety of data transmission tasks. The chief use has been to transmit the "closed captioning" that assists the hearing impaired and those who have pressed the mute button. (Closed captioning uses only one line of the 24 that are available for data transmission.) In sending such data, broadcast stations become at least for that tiny fraction of a second unregulated broadband information providers. The VBI could be used in conjunction with the Internet to establish a partnership between local TV channels and Web pages (see Chapter 17). The same technology that is used to insert data into the VBI could be used to insert data into all 525 lines of the video frame during periods when the broadcaster is dark. The VBI is a small but powerful way for local TV stations to partake in the information revolution.

The Future

The "ancient history" of broadcast radio and broadcast television has much to teach us about the future. First, the business models and strategies that worked for radio and television are likely to be important, even with changes in technology, because new media have similar economic characteristics. Second, the regulatory history of broadcast radio and television offers clues to the central role the federal government will play in the development of new forms of video delivery. Both television and a large part of the Internet (especially access lines and any radio-based transmissions) continue to be subject to federal regulation. More of these industries may be regulated, especially if it suits the interests of a powerful lobby.

Chapter 15 will discuss the future of digital TV broadcasting. One day, every present-day broadcaster may have control of digital capacity sufficient for six or more conventional TV signals. Without a doubt, the government will keep its hand in the works. There is a strong likelihood that broadcasters will be required to continue to supply "free" over-the-air television programs, even though they must compete against those who are permitted to charge. Furthermore, 6 megahertz probably will be insufficient bandwidth to permit offerings competitive with nonbroadcast entities. Nevertheless, broadcasters may be prevented from combining in joint ventures or otherwise achieving efficient scale—prevented by rules designed for a different time, perpetuated by interest group pressure mounted by their commercial rivals.

Conclusion

Television is one of the most spectacularly successful commercial products of all time. This success derives in major part from television's ability to serve unmet consumer demands. Success also arose from the fact that the business model (radio) and regulatory structure for television already existed.

The decision to limit broadcasters and networks to a single channel apiece limited the ability of broadcasters to offer program packages that would be appealing to viewers. Newer media such as cable and satellite broadcasters are able to package channels in a coordinated way not available to traditional broadcasters.

Remote control devices made fundamental changes in the demand for television. Viewers could more readily search for preferred options. But this made television less satisfactory as a passive entertainment medium.

An unused segment of each broadcast channel called the Vertical Blanking Interval can be used to transmit data at the same time as regular programming. This may give broadcasters the capability to provide an Internet-content transmission path.

CHAPTER SEVEN

THE EVOLUTION OF CABLE TELEVISION

Television stations are located near big cities, and their transmissions are aimed at population densities. People living in rural areas, or surrounded by mountains, cannot get good over-the-air TV reception. Nothing better illustrates the strength of viewers' demand—willingness to pay—for television than the growth of the cable television industry, which for many years did nothing more than improve reception of over-the-air broadcast channels, for a fee.

Television by Wire

The earliest cable systems and the first commercial television broadcasts appeared at the same time—the late 1940s. Many of the early systems were jury-rigged—literally nothing more than "extended antennas," as the Supreme Court would later characterize them. A TV

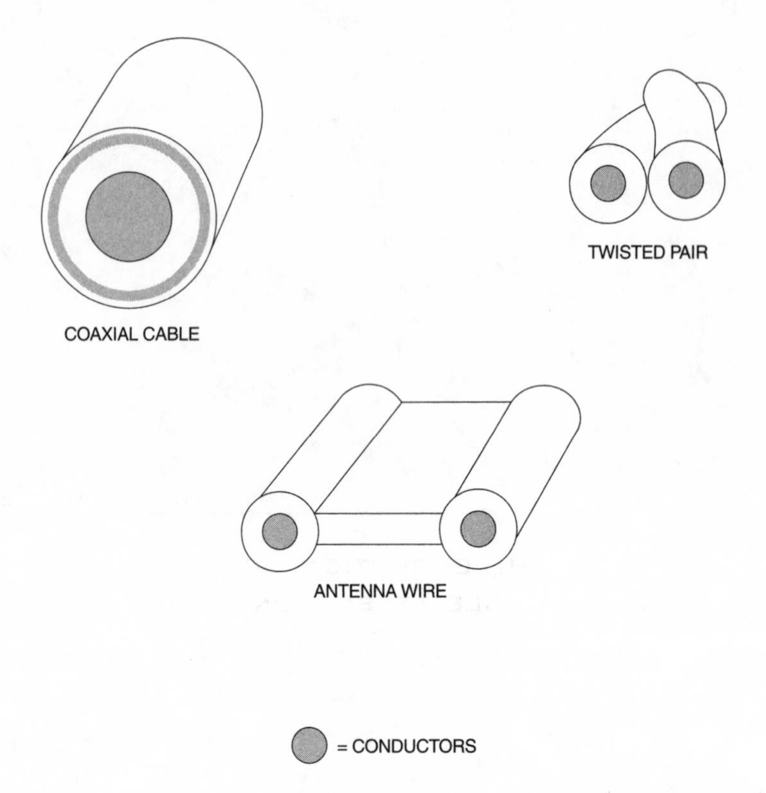

Figure 7.1 Transmission cables: coaxial, telephone, and antenna

set retailer or repair shop would install an antenna on a nearby hill, run a wire to the store, and then later to individual households, and start charging by the month. Because the service was of interest only to those few households that could not receive regular signals, it was a very small business at first. Cable service was further restricted to rural households located in clusters large enough to make a cable system economically feasible.

The technology of the early cable systems was not much more complicated than the technology of a rooftop TV antenna. At the time, the most efficient way to transmit video signals for distances of a few miles was coaxial cable—a broadband transmission line consisting of two cylindrical copper conductors arranged concentrically, as illustrated at the top left of Figure 7.1. The very first cable system used flat antenna wire, illustrated at the bottom of Figure 7.1, but the second system and all subsequent ones have used coaxial cable, from which the industry took its name. (For comparison, an ordinary telephone wire or "twisted pair" is also shown in the figure.)

Even in a household that uses a rooftop antenna, if the TV antenna is to provide signals to multiple TV sets it may be useful to install an amplifier on the cable from the antenna to the set. An early cable system had an antenna, a network of splitters and coaxial cables, and enough amplifiers to ensure that each subscriber's signal was adequate for good reception.

A complete coaxial cable system is depicted in Figure 7.2. Program sources include local TV stations, distant stations, satellite programming, and local origination. Trunk cables carry the signal to the distribution plant. Limits on the number of amplifiers that can be placed on a line dictate the relationship between trunks and distribution or feeder cables. Each subscriber is served by a drop line that runs from a tap on the distribution cable to the residence. Inside wiring (which is also coaxial cable) carries the signal to a converter box (which may also descramble premium channels to which the household subscribes). The signal then typically passes through a VCR to the television set. There must be as many converter boxes as there are televisions in the house, if all are to receive cable service.

The number of channels that can be carried on a coaxial cable system is a function of the design capacity and power of the amplifiers and the design of the converter box that sits on top of each subscriber's TV set. The earliest systems had cheap amplifiers and no converter boxes; they had a bandwidth of 100 megahertz or less and offered twelve or fewer channels, enough for all local stations and some local origination or distant signals. Later, as additional cable services became available, the bandwidth and power of amplifiers were increased, set-top converters were introduced, and either fiber-optic or dual cable links were installed. These systems accommodated the rapidly increasing number of cable networks. Table 7.1 shows typical bandwidths and channel capacities for cable systems.

The most advanced cable systems thus have about 1 gigahertz of bandwidth, roughly the equivalent of 1.5 to 2 gigabits per second. The new "wireless cable" systems (local multichannel distribution service, or LMDS) also have approximately this bandwidth—no accident, because the FCC was attempting to create an effective wireless competitor to cable.

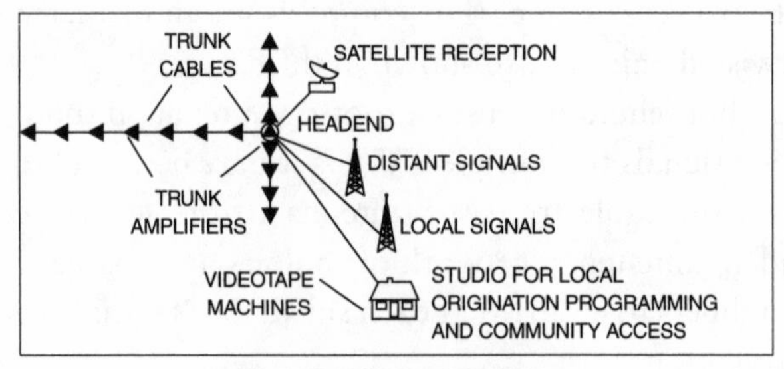

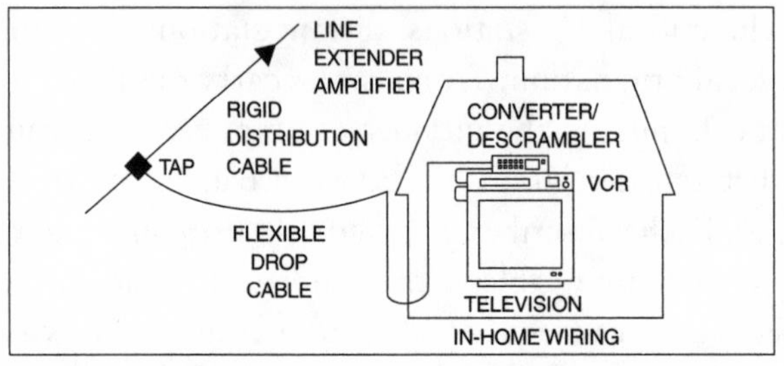

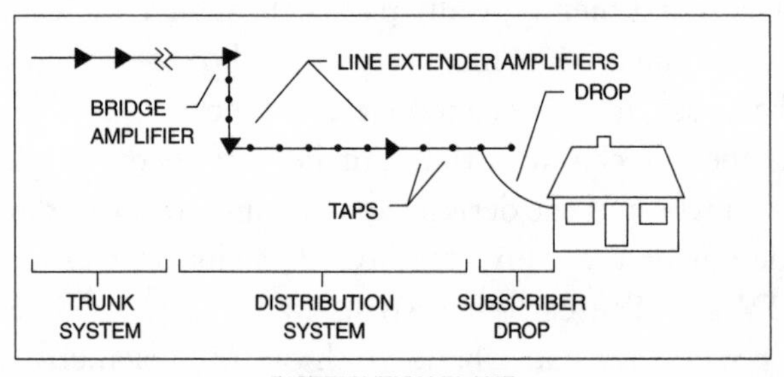

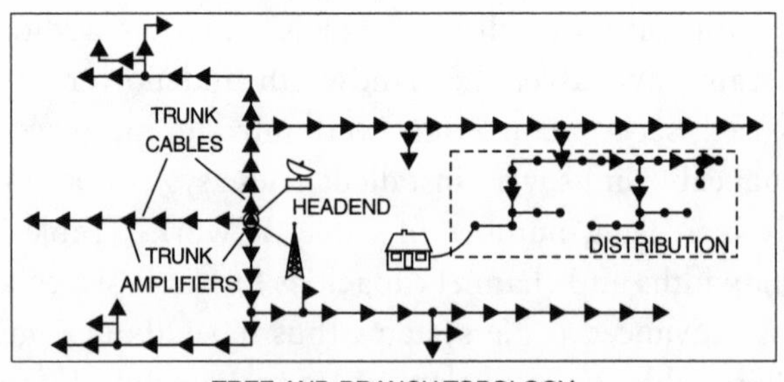

Figure 7.2 Cable system architecture (Ciciora, 1995; used by permission)

Table 7.1 Cable system bandwidth and channel capacity

Size of system	Bandwidth (megahertz)	Operating frequencies (range in megahertz)	Number of channels (before compression)
Small	170	50–220	12–22 (single coax)
Small	220	50–270	30 (single coax)
Medium	280	50–330	40 (single coax)
Medium	350	50–400	52 (single coax)/ 104 (dual coax)
Large	400	50–450	60 (single coax)/ 120 (dual coax)
Large	500	50–550	80 (single coax)
Large	700	50–750	110 (single coax)
Large	950	50–1000	150 (single coax)

The traditional cable system (Figure 7.2) had a tree-and-branch structure, with coaxial cable from one end to the other. In contrast, modern cable systems are a hybrid of optical fiber and coaxial cable (HFC). (A diagram appears below in Figure 7.6.) In this respect, cable systems and local telephone systems use similar technologies. Optical-fiber trunks transport data to nodes serving neighborhoods of perhaps two thousand households. Coaxial cable (or, in the case of telephone systems, twisted copper wire) carries the data to individual households.

There are two main economic reasons for this structure. First, although fiber is less expensive than either coaxial cable or twisted pairs, the equipment needed to translate optical signals into electrical ones that can be understood by current TV sets and telephones is quite expensive. This is not computer equipment, and its price may not fall as rapidly as the price of computer equipment. By translating the optical signals for a whole neighborhood at one node (where fiber-optic transmission connects with coaxial cable transmission), the cost of the equipment is spread over many subscribers. Therefore, the system as a whole is much less expensive than if the equipment had to be installed in every household. Second, while fiber is cheap, installing it is not. The farther downstream one goes from the cable headend (or telephone switch) toward the subscriber, the more miles of fiber must be installed. For a cable system, only about 12 percent of total system mileage is trunks. The rest is local distribution (38 percent) and drop lines (50 percent). The key economic variable in designing an HFC system is the ratio of nodes to subscribers. At one extreme, a node can be placed in each home, and at the other extreme there can be just one node, at the headend. An HFC upgrade costs around $300 per home passed (the cost of the upgrade divided by the number of homes it could serve), and of course much more per subscriber. There are cost, capacity, and performance tradeoffs between these extremes.

Why use fiber at all, if the cable coming into the home is still coaxial? Fiber is a cheaper way of transmitting large numbers of channels over the relevant distances. In other words, fiber does nothing that coaxial cable and amplifiers could not do; it just does it more cheaply per bit per second, especially at high bit rates. As noted above, the capacity of a coaxial cable is around 1 gigahertz, equivalent to perhaps 2 gigabits per second. The capacity of a single strand of optical fiber is 25 terabits (trillion bits) per second for each of the three frequency bands or wavelengths it can support. This is enough to carry all of the peak-hour telephone traffic in the United States (about 1 terabit per second) on a single strand. This capacity, however, is theoretical. It is not achievable in practice for various reasons, including the slower speed of electronic (nonoptical) equipment to which the fiber is connected. But the practical capacity of fiber is still measured in terahertz. Optical-fiber capacity is increasing, as new methods of modulation and multiplexing are developed.

Given fiber's enormous channel capacity, why not take it all the way to the home as provision for future demands for more capacity? This is the key question. Cable television and telephone company executives have given two answers. First, they are not sure there will be sufficient demand for these unknown new services to pay for any capacity increase, much less the enormous one that fiber permits. Second, regardless of the demand, they are not sure what will be the cheapest way to deliver the capacity. Fiber is a "hardware solution" to the capacity problem. To install fiber to the home may cost $1,500 or more per subscriber, a number roughly as large as the current capital asset value of a cable or telephone subscription. (According to some estimates, about 70 percent of this amount is "inside" equipment at

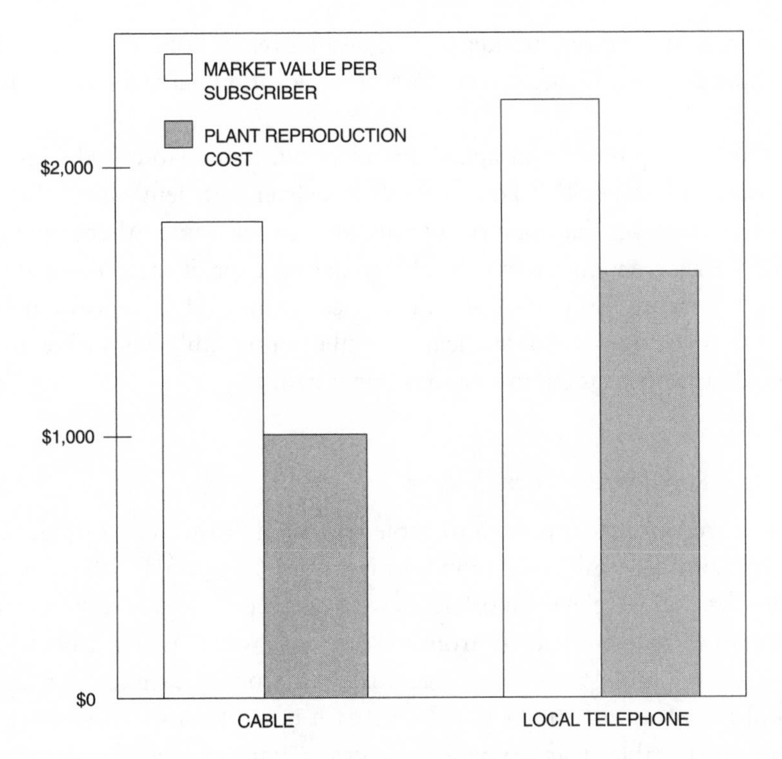

Figure 7.3 Capital values for cable and telephone subscribers (MacAvoy, 1990)

the ends of the fiber, while only 30 percent is the cost of actually installing the conduit, a labor-intensive process. There is every chance that the prices of the terminal equipment, like those of other electronic products, will decline.)

Figure 7.3 provides ballpark estimates of the market value and reproduction costs, per subscriber, of residential cable and telephone service. (Reproduction cost is the cost of replacing an older facility with a new one of the same type.) These investments, of course, are already in place. Either would have to be upgraded to provide interactive video services. The ratio of reproduction cost to market value (called "Tobin's *q*") is regarded by some economists as an indication of market power. The idea is that entry (or regulation) should drive market values into equality with reproduction cost. By this measure, according to Figure 7.3, both cable companies and telephone companies have substantial market power, despite regulation. The issue is a complicated one, however, and further exploration here would take us too far afield.

Coding equipment attached to telephone lines (so-called digital subscriber lines, or DSL) may be able to deliver sufficient capacity at a cost much lower than that of running fiber to the home. Alternatively, so-called cable modems may be able to deliver capacity at a reasonable price. Both are "software" solutions because they rely heavily on new methods to compress data cheaply. A fiber-optic cable has such enormous capacity that compression is unnecessary.

On the Regulation Roller Coaster

The government's approach to cable television illustrates its predilection to protect established economic interests. Compared with rooftop antennas, cable television offers viewers greater choices and therefore "siphons" audiences away from local broadcasters. When it became clear in the mid-1960s that cable posed an economic threat to many local broadcasters, the FCC did all that it could to slow, or stop, the growth of cable. Cable operators were initially denied the right to import TV channels from outside a market; they were forbidden to offer pay programming that included series episodes, or movies more than two years old, or sports events that had been on conventional television.

More recently the commission (and then the Congress) required cable systems to carry all local broadcast stations without charge, regardless of the merits of their programming. These "must carry" rules were later upheld (under a First Amendment challenge) by a Supreme Court plurality. The local broadcast station has it made: either it can withhold its signal from the local cable operator unless the cable operator pays for it, or the broadcaster can force the cable operator to carry the signal gratis. Special exceptions to the copyright law are made in either case. It is difficult to imagine a more one-sided arrangement, and it is no accident that this broadcaster-sponsored provision was included in the same 1992 Cable Act that reregulated cable rates in the heat of the 1992 election.

As the growth in cable channel capacity illustrates, video communication is just a matter of capacity, and the economic viability of the service is just a matter of cost-per-unit capacity. The broadcast airwaves have essentially unlimited capacity, but regulators chose to restrict the amount available for TV broadcasting. It also decided to set the technical standards governing the imprinting of video images on the airwaves. These decisions have limited consumers' video choices. In addition, until the 1970s rules barring broadcasters from charging viewers guaranteed an excess demand for television programs. Cable systems also have essentially unlimited capacity, but the amount actually employed is set by the supply and demand for their programming, rather than by the government.

Cable satisfied the excess demand for television programming created by the FCC's restrictive economic policies. If the commission had made adequate provision for broadcasting in its allocation of airwaves, or had set less rigidly controlled modulation standards, the cable television industry might not have sprung to life, or at least not outside rural hamlets. Conventional or near-conventional use of the airwaves may have been the most cost-effective way to deliver 120 channels of TV service to people's homes. This is a staggering possibility. The cable television industry has invested more than $75 billion in cable plant. Yet this fortune may have been spent, from a social perspective, for no good reason. It was largely a wasteful effort to permit consumers to get around misguided federal constraints on program supply.

Alternatively, if the government had charged broadcasters an efficient price for use of the spectrum, it might have turned out that broadcasting was inefficient everywhere except in rural hamlets, and that cable was the most efficient urban delivery medium. (In other words, in this scenario broadcasters would not have been able to outbid cable companies for programming and advertising.) If this view is accurate, what is interesting is that cable succeeded even though the government did not supply cable channels for free.

Here is how cable came to satisfy the excess demand for video programming, starting in the 1960s. Once in business as "extended antennas," cable operators began looking around for new sources of revenue. Improving rural TV reception was, after all, a limited market. They had two choices: to start wiring isolated ranches and homesteads, or to aim instead at larger population centers. The trouble with wiring isolated homesteads is that it costs a lot to string the needed wire, especially in relation to the potential revenue. And yet the trouble with the larger population centers, aside from those like Manhattan with tall buildings that blocked reception, was that TV viewers had tolerably good over-the-air reception already, and hence little demand for cable.

It became clear that to enhance the demand for cable service in areas where over-the-air reception was not impaired, new services had to be added. In the early days there were two categories of new cable services. The first was "local origination." A TV camera would be aimed at a clock, or a weather instrument, or a news ticker. At the time, this was novel because such information had not previously been continuously available on television. The second was distant signals. Distant signals were out-of-town TV stations that had programming (chiefly sports, old movies, and off-network reruns) not available on the local stations. To get the distant signals, the cable operators at first simply raised the height of their antennas. Later, they set up antennas near the distant cities and used microwave relay towers over considerable distances to import the desired stations. For example, the Denver stations were imported throughout the Rocky Mountain region.

Distant signals were essential to the growth of cable until the late 1970s. Yet they were also extremely controversial in Washington. From the FCC perspective, the distant signals contained copyrighted material—intellectual property in the form of programs—to which the cable operator had no clear right and for which it paid nothing. Regulators and broadcasters thought that to the extent that imported programming duplicated programming already available in a local market, its value to the local broadcaster fell, forcing copyright owners to charge less for it. What is more, the imported signals fragmented the local broadcast audience, reducing advertising revenues of local stations in markets with cable penetration. Both broadcasters and program producers had reason to be concerned, and they complained loudly to the FCC and in court.

The copyright issues were resolved first. The Supreme Court in 1968 held that cable operators violated no one's property rights when they simply put up an antenna (however tall) to improve reception. The Court reasoned that a householder putting an antenna on her roof was not liable for copyright payments on the transmission that took place between the roof and the TV set, and that the cable TV system was a mere extension of that rooftop antenna. A few years later the Court reached a similar conclusion with respect to imported distant signals: the antenna was just being further elongated.

The distant signal issues occupied the FCC for nearly two decades. When broadcasters began to complain in the late 1950s about competition from cable systems, the commission brushed them aside. It ruled that it had no jurisdiction over this minor rural issue. By the early 1960s, however, the broadcasters' protests became shrill. The commission reversed field and asserted its jurisdiction over cable systems. It decided it had to limit the importation of distant signals because of the economic harm to broadcasters. In the late 1960s, against the backdrop of Court decisions absolving cable operators of copyright liability, the commission acted to freeze all new distant signals.

The government's official explanation for its distant signal restrictions was the need to protect local UHF television stations, whose audiences might be reduced by imported signals. There were three problems with this explanation. First, it was paternalistic. It denied viewers the option to see programming they preferred, and it forced them either to view local programs they valued less, or not to view TV at all. Viewers should support localism, it was thought, and regulators would see that they did. Second, the explanation was dead wrong on the facts. Cable systems promoted the growth of local UHF stations, by improving reception, far more than the imported distant signals hurt the stations. Finally, the explanation was a lie. The real reason the commission acted as it did was not because of the little guys, the fledgling UHF stations, but because of the economic worries and political pressure of the stronger broadcasters—the VHF network affiliates. Moreover, in order to ensure the continued economic security of its major broadcast licensees, the commission went further, making it virtually impossible for cable systems to turn to pay-TV as an alternative source of demand.

The pay-TV or "antisiphoning" rules adopted in the 1960s seem laughable today in their baldly anticonsumer, protectionist tone. Cable operators could not offer pay-TV (per-channel or per-program) service that included (a) sports programming that had been on free television within the past four years, (b) series programs, or (c) movies more than two or less than six years old. The idea was to keep cable operators from competing to buy the programming then purchased by major networks and broadcasters, and thus driving up its price. The rationale was to protect viewers' supposed interest in having their free programs (with commercials) saved from siphoning to pay-TV (without commercials). The District of Columbia Circuit Court struck down these rules in 1977.

Enlightenment and reform slowly overcame the cable television regulatory process, starting in 1972. By 1979 the FCC had repealed essentially all the distant signal restrictions as part of the general deregulation movement then in fashion. Meanwhile, cable growth became exponential, spurred not by distant signals but by pay-TV in the form of the Home Box Office (HBO) movie channel and by new basic service networks. HBO was a standout attraction for cable. Indeed, many people referred to cable in those days as "HBO" rather than "cable." HBO, transmitted by C-band satellites (see Chapter 8), was the early key to cable penetration of major urban areas, where viewers already had plenty of good over-the-air choices. And it was litigation sponsored by HBO that led to the federal court decision striking down the FCC's pay-TV restrictions.

The late 1970s were the years when the specialized cable channel was born. Program suppliers such as Ted Turner, with his decision to create a "superstation" from WTBS, his Atlanta TV station, discovered a market in selling cable channels to cable operators who used them to expand the demand for cable service among metropolitan and suburban viewers. Later, as audiences grew, these channels also gained from the sale of advertising. By 1980 there were nearly thirty nonbroadcast cable channels, including HBO. By 1990 there were more than seventy-five nonbroadcast cable channels available for sale to cable system operators. With more channels to offer, cable operators were able to raise the price of cable service (though not the price per channel). Unfortunately for the cable industry, rising prices led to public protests and calls for price regulation, which in an election year (1992) two-thirds majorities of both houses of Congress were willing to provide.

It Takes Two to Tango

More than relaxation of the government's anticable rules was required to make cable a viable economic proposition in metropolitan areas.

Although consumers appeared willing to pay for additional video channels and firms found ways to assemble the necessary programming, there was no effective way to get the programming to the cable systems.

Prior to 1974, AT&T microwave facilities distributed network television programming to TV stations. Nonnetwork programming, such as old movies shown by local stations, was "bicycled"—sent by mail or messenger. A typical TV broadcast network had about 220 local affiliated stations around the country. Supplying them with programming brought in an AT&T microwave bill in the tens of millions of dollars per year, per network. On a per-hour basis these rates were even higher for part-time users. But there were some 3,700 cable systems in 1974. To connect them with AT&T's television relay service would have been far too expensive. Indeed, to import distant signals, cable operators built their own microwave systems.

The solution was the geosynchronous communication satellite (GEO) discussed in the next chapter. In 1974 Hughes constructed and RCA purchased and launched the first such satellite to serve the United States. A GEO is like a broadcast station with a very tall antenna. Everyone "illuminated" by its beam can receive its signal; the number of receiving stations has no direct effect on the cost of the satellite. (Nearly all of the energy emitted by a broadcast satellite is wasted—absorbed by water, foliage, and other nonantenna objects.) This public-good aspect of a satellite meant that it was perfect for distributing TV signals to thousands of cable systems. The effect was to drastically reduce the cost of interconnection for new cable networks. Satellites, as much as the changes in the regulations or the Supreme Court ruling in the HBO case, were responsible for the success of the cable industry.

How Cable Systems Make Money

A cable television system requires a considerable up-front investment to build the plant and then on-going expenditures for maintenance, customer connects and disconnects, and programming. The system will be successful if the discounted present value of the net cash flow over its lifetime exceeds the initial investment cost.

Cable television systems seldom show an accounting profit, even though they are sold and resold for $1,500–$2,000 per subscriber. The reason is that accounting depreciation, appearing as an expense on the income statement, typically wipes out any profit. Firms with no profit pay no taxes, so cable systems pay no taxes. Far from depreciating, however, the value of cable systems over the years has generally appreciated. So a person who purchased a cable system for $500 per subscriber in 1980 and then made accounting losses while operating the system for ten years, might sell the system for $1,000 per subscriber in 1990. The new owner would start out with a new tax "basis" of $1,000 for future depreciation, giving rise to even larger annual tax write-offs.

Cable is a cash-flow business, much like commercial real estate. People pay good money for the common stock of cable companies that have never made a profit or declared a dividend. Typically, a large portion of the capitalization of cable companies is in the form of bank debt, making the companies highly leveraged. As we shall later see, this means that the companies and their stockholders are especially vulnerable to competitive threats, such as broadcast satellites.

Financial development in the cable industry followed a pattern. Multiple system operators (MSOs)—such as Telecommunications Inc. (TCI), Time Warner, Comcast, and Cox—accumulated noncontiguous cable systems through purchase. The largest MSOs own thousands of systems scattered around the country. The MSOs can operate these systems more efficiently than can individual owners, presumably, because of lower cost of capital, superior managerial expertise, and perhaps lower programming and equipment costs from suppliers' bulk discounts. Historically, at least as much money has been made from buying, depreciating, and selling cable systems as from operating them. Also, until recently, constructing and running a cable system required little sophisticated technical expertise. Large MSOs were more like financial companies than communication companies.

Unfortunately for cable MSOs in today's competitive world, things have changed. First, making the right guess about the future of the business requires a fundamental understanding of highly sophisticated and even exotic new technologies. Second, most MSO cable systems are too scattered and too local to offer the most effective competition to other communication media. The geographic territory served by a single broadcast station may have dozens of cable systems. The territory of a Bell Telephone operating company may have thousands. The area served by Hughes's DirecTV satellite (the whole continental United States) has more than 11,000 cable systems. Thus Hughes can economically promote its service in the national media, while no cable system or MSO can do the same.

In recent years cable systems serving individual metropolitan areas have begun to consolidate through swaps among the MSOs. They have a long way to go. Only about half of all subscribers are served by clusters of cable systems with more than 100,000 subscribers. Even when they are integrated through ownership, it is challenging to integrate contiguous local cable systems physically or in a marketing sense because of technical differences in system characteristics (age, number of channels) and especially variations in local franchise requirements.

The geographic pattern of cable system development is based on state laws authorizing local franchising (licensing) of utilities that make use of streets and other public facilities. In many states individual towns and villages each have the right to franchise any number of cable operators—but usually license only one. The franchise competitions that took place in the 1960s and 1970s had no features ensuring that a given MSO would tend to win the franchises in broad contiguous areas. One reason for this was the lack of interest of local authorities in efficient operation, low costs, and low prices. Instead, city mothers and fathers focused on bells and whistles—for example, channels for the school system, advanced technologies, and neighborhood studios. (Sometimes, their focus was on borderline or outright bribes; at least one prominent cable executive went to jail.) The cable systems were built as individual units. Until it came time to worry about competing with other media that might enjoy the same cost or lower cost per bit as cable, neither cable companies nor local authorities saw much to be gained from regional consolidation.

Typically, cable operators pay their network suppliers on a per-subscriber basis. In some cases, such as home shopping channels, the network pays the cable operator. Figure 7.4 shows a simplified view of some of the economic relationships in the cable industry. (Various important features, such as the relationship between cable and local broadcasters, have been omitted.) This pattern is similar to the arrangements between broadcast networks and their affiliates, in which there is a flow of cash to the affiliates (see Figure 6.2). One very impor-

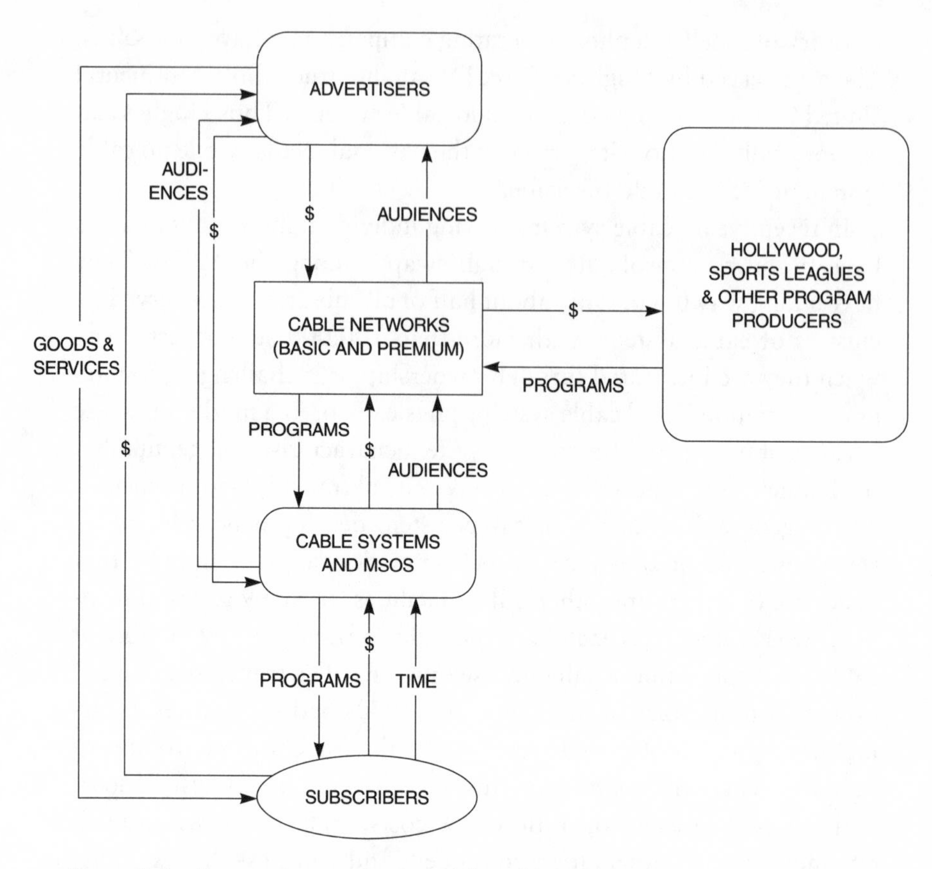

Figure 7.4 Economic organization of the cable industry

tant difference is the flow of money upstream from subscribers to cable operators to networks and program producers. It is this flow that permits cable networks to offer attractive programming in competition with the broadcast networks even though their audiences, and hence their advertising revenues, are much smaller.

Table 7.2 shows a typical cable system's operating statement on a per-subscriber basis. Revenue is derived chiefly from advertising, programming (extended basic service), basic service, pay-per-view service, and installation charges. Noteworthy is the relatively small share of expenses devoted to programming. The depreciation expense, as already noted, is partly an accounting fiction that acts as a tax shelter.

Table 7.2 Cable revenues and operating expenses, 1995 (dollars per subscriber per month)

Revenue		Operating expenses	
Advertising	5.75	Payroll	8.90
Programming	9.05	Programming	12.64
Basic service	23.40	Depreciation	6.75
PPV service	8.47	Lease and rent	1.00
Installation	.72	Advertising	1.14
Other	1.87	Other	8.47
Total	49.26	Total	38.90
		Contribution	10.36

Note: For 58 million subscribers.

Two risks faced cable television operators in the early days. First, it was not known whether there would be a sufficient demand for cable service outside of rural areas to return a profit on the investment required. Second, it was not known how long restrictive government policies would continue. For more than twenty years the federal government had attempted to retard the growth of cable television in order to safeguard the profits of TV stations and networks.

Faced with these risks (and others), firms invested more than $75 billion in cable development. Today, 97 percent of all households can subscribe to cable if they choose because cable operators have paid to lay cables down almost every street. Cable operators made this investment in the expectation of earning a profit. As it turned out, they were successful. The government did reduce its pressure on cable, starting around 1978, and when new program services like HBO and CNN became available, suburban consumers flocked to sign up for cable service. Cable operators began to reap the profits they had hoped for at the time of their initial investment.

Unfortunately for the cable industry, economic success (and serious service quality problems) attracted the attention of elected officials, who accused cable operators of enjoying an economic monopoly and in many cases offering inferior service. As a consequence, the federal government in 1992 imposed on cable operators price and profit controls. The 1996 Telecommunications Act lifted some of these controls beginning in 1999, or when effective competition is in place.

Notwithstanding cable's own role in bringing regulation upon itself, the experience of the industry demonstrates to every entrepreneur considering investments in new communication technologies that positive outcomes may be truncated by politicians. Taking on a risk that produces too much success will be penalized, not rewarded. Accordingly, there will be some investments that cannot be justified, and some technologies that will not be developed. The prospect that great success will be penalized by the government can only stifle innovation.

Suppose it would cost around $600 per subscriber to build a cable system in a given community. Why would anyone be willing to pay $1,500 or more per subscriber to buy such a system, as often happens? After all, if the system is worth $1,500 per subscriber that must mean that the risk-adjusted discounted present value of expected future profits (cash flow) is $1,500. But if that is true, and if it costs only $600 per subscriber to construct a system, why not just construct a system in parallel with the one offered for sale at $1,500? In other words, in a market that was open to competitive entry, profits would be driven down so that the system would be worth only about $600.

The disparity between what a cable system costs to build and what it is worth has led many to conclude that the cable operators must have monopoly power. Otherwise, the $900 "economic rent" implied by the example would be wiped out by competition, driving prices toward costs. There are, however, other explanations. One is that a new system can expect to have startup losses that must be capitalized just like construction costs. These have to be added to the $600. Another is that at least part of the discrepancy is attributable to risk taking. As highly leveraged (debt-financed) firms, cable systems are especially vulnerable to competitive risks and regulatory and tax law changes.

Until recently, the most direct form of competition to cable operators was an "overbuild"—two systems operating side by side, as in Allentown, Pennsylvania. But overbuilds are rare. Nowadays a more general threat looms: actual competition from satellites and potential competition from telephone companies. Given that cable plants are not useful for any other purpose, cable operators could face huge capital losses if the new competition proves to be more efficient or more attractive in delivering programs to viewers.

There are additional explanations of the hypothetical $1,500 figure. For example, at least part of that amount is attributable to the value of the tax shelter associated with the new, stepped-up asset value that forms the basis for future depreciation. This may be an economic rent, but it is attributable to tax laws rather than to market power. Finally, the very notion of competition is difficult to apply to this situation because the fixed costs of a cable system are so large. That is, if a new entrant "overbuilt" an earlier cable system, there would be a high risk of a price war that could result in elimination of the new entrant's investment. Therefore, it would be foolish for a new entrant to count on profits sufficient to support even a $600 investment, much less a $1,500 investment.

It Keeps Getting Better

From society's point of view the performance of an industry is measured by improvements in its products, growth in output, and reductions in costs and prices. The cable industry in the 1980s made astonishing progress in all three respects. As illustrated in the four panels of Figure 7.5, product quality as measured by program expenditure rose rapidly, output as measured by channels per system and by number of subscribers also increased, and prices, measured on a per-channel basis, fell. Few industries performed this well in the same period.

These achievements were sufficient to bring cable penetration to nearly 60 percent of all TV households by 1990. But there cable stalled. The industry has been able to increase penetration by no more than 6 or 7 percentage points in the 1990s. (There is thought to be around 5–10 percent additional penetration that is stolen.)

Always a Price to Pay

Even though the quality and quantity of programming improved, subscriptions increased, and real unit prices fell throughout the 1980s, cable subscribers became increasingly alienated from their suppliers. First, many cable operators were slow in responding to repair re-

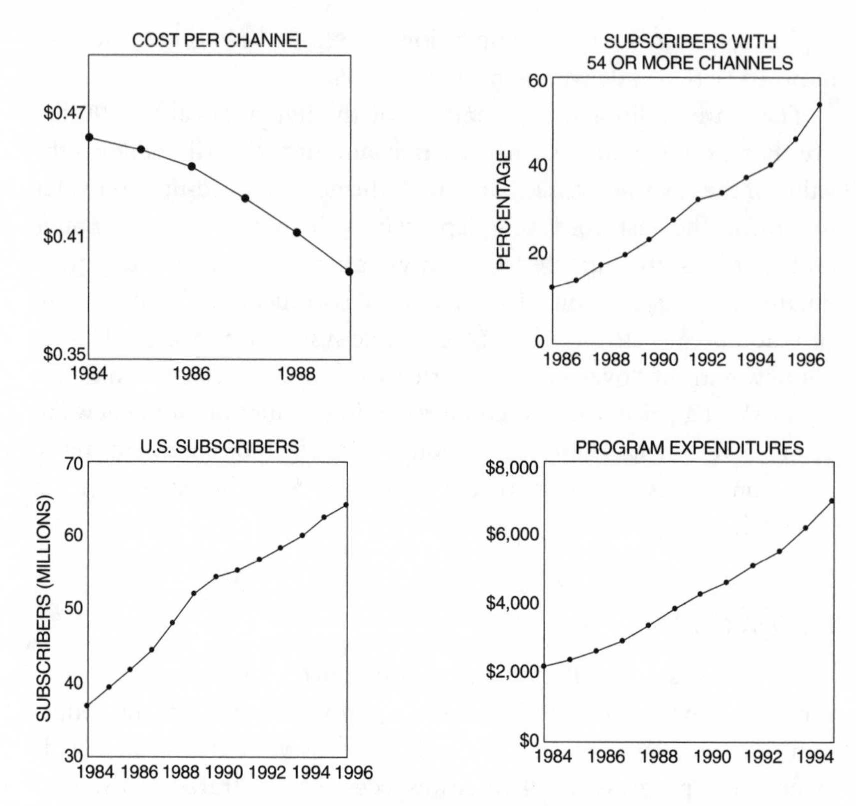

Figure 7.5 Economic performance of the cable industry during period of deregulation (U.S. Federal Communications Commission–General Accounting Office, 1990)

quests, and some provided less than reliable service. Second, even though the price per channel was falling, the overall monthly fee for service was rising.

Cable operators add new channels to attract new subscribers and to retain marginal old ones. Subscribers content with the basic or older services may have little use for the newer ones and may therefore perceive the price for their channels as going up. Due in part to the negative public image of cable systems, and due in great measure to inept public relations by certain MSOs, Congress repeatedly threatened and in 1992 actually imposed federal rate regulation.

Economic rate regulation of an industry, however justified, invariably leads to inefficiency and consumer losses. It is difficult to discern a sound policy basis for the Cable Act of 1992. Cable television service is hardly a basic necessity, like electricity, water, and perhaps telephone service. Moreover, the United States had just completed a period of deregulation in which the prices of bus service, railroads, trucks, airlines, and certain energy products were left open for market determination. Indeed, in 1984 the Congress had taken away from the states and cities the power to regulate cable rates.

Congress delegated to the FCC the power to fashion a rate regulation scheme for cable, subject to various broad (and internally inconsistent) guidelines. The commission, at the end of the Bush administration, came at its task somewhat reluctantly. From the FCC's point of view, the obligation to regulate the rates of 11,000 or more cable systems was daunting. In order to avoid a case-by-case determination, the commission sought formulas and rules of thumb.

One issue was whether to treat "basic" cable service differently from "expanded" service. Basic service is defined (by statute) as local broadcast signals and public service channels. Expanded service, generally offered in one or more tiers, is composed of cable networks such as CNN, The Learning Channel, and The Discovery Channel. (A third category consists of premium channels, such as HBO.) The 1992 act could be interpreted to require relatively strict controls on basic service, light controls on extended basic service, and no controls on premium channels. Cable operators viewed these constraints as a challenge. Quite naturally, they tried to find a way around the restrictions, to adjust matters so that the adverse impact of the regulations was minimized. Attempts by regulators to treat different kinds of cable service differently led to the pilgrimage of programming into the service categories least constrained by FCC regulations—premium channels and newly added channels and packages.

This illustrates a general principle of economic regulation, the "balloon problem." Press the balloon anywhere, and it pops out someplace else. Consider the effects of the FCC's Fairness Doctrine. The Fairness Doctrine required broadcasters to air all sides of controversial issues. But to do so was unprofitable, so broadcasters simply avoided controversy altogether. After 1987, when the doctrine was repealed, there was an explosion of controversial material in the broadcast media. Finally, all this parallels the notion, discussed above, that the cable industry itself can be viewed as a "way around" the restrictive policies that prevented consumers from expressing their economic demand for more and better programs.

The FCC ultimately adopted an elaborate set of formulas designed to roll back the rates of cable systems whose rates were well above average after correcting for various system characteristics. The cable industry suggested the framework of this scheme, though not the actual numbers. But shortly after the commission acted, the Clinton administration's new FCC chief, Reed Hundt, came into office. Hundt was highly critical of the commission's work; his political mentor, Vice President Al Gore, had been the chief critic of the cable industry while a member of the Senate. Though retaining some of the commission's previous framework, Hundt imposed a 16 percent price reduction on nearly all cable systems, regardless of their prior pricing behavior.

One result of Hundt's heavy-handed initial approach was that a significant number of cable systems took refuge in their constitutional right to offer a full cost of service justification for their fees. Adjudication of these proceedings would have overwhelmed the resources of the commission. In the end, Hundt entered into so-called social contract settlements with many of the MSOs. These typically called for modest price reductions combined with some public relations gimmick, such as a small rebate, to make it appear that the cable operator was giving up more than it was.

Once again, federal regulation of cable television rates is seen as a mistake and an embarrassment. The Telecommunications Act of 1996 provides for the phasing out of all cable rate regulation, except for the "basic" tier, after 1999. But there is no guarantee that consumer complaints about cable rate increases will not call forth renewed regulatory legislation.

Cable Lessons

The role of government is central to the success or failure of the cable industry. This has been demonstrated repeatedly. The industry has gone through several difficult cycles of being first in and then out of political fashion. The cable industry is not unique among communication industries in this respect, and there is no prospect that this will change in the future. No communication business touched by federal regulation can afford to ignore the possibility that Washington will intervene in its affairs. For that matter, no communication business should ignore the opportunity to seek benefits for itself from regulators, or to seek to impose costs on its competitors.

One necessary condition of the cable industry's success was a serendipitous technological development in an entirely different industry, the geosynchronous communication satellite, or GEO. No communication business can succeed looking only at developments in its own technology; it must also consider those in other businesses that might have positive or adverse external effects.

In calculating expected returns, investors in risky new communication technologies should not assume that they will be permitted to keep all of the rewards of great success. The cable industry will continue to face enormous competitive and financial risks. The threat of renewed rate regulation may truncate profit potential and make it less likely that the cable industry will be able to raise the capital needed to prevail in the coming competitive battles to supply digital services to consumers.

Misnamed Modems

The future of cable television, while it may not have much to do with the Internet, will certainly intersect digital television technology, cousin to the Internet. An explicit premise of the Telecommunications Act of 1996 is "convergence"—telephone companies will be offering video service and cable companies will be offering telephone service. Competing with the phone companies, cable companies can provide Internet access at bandwidths suitable for interactive video. A flurry of cable industry announcements—very premature as it turned out—in 1996 concerned the proposed acquisition of millions of "cable modems," as they are sometimes called. As of 1998 there were a few dozen experimental cable systems with a total of about ten thousand cable modems in service in the United States. Vendors are said to have orders for the delivery of hundreds of thousands, some of them for devices still under development.

In late 1996 Scientific Atlanta announced a cable "modem" that would offer one-way service through the cable system with a return data path through the telephone system. To that end, the digital cable modem has a built-in analog telephone modem. The cable modem would be capable of downloading data at up to 1.2 megabits per second. (This is probably not enough for broadcast-quality video, ironically, on a video medium, but it is enough for highly compressed video, and probably offers capacity similar to asymmetrical digital subscriber lines (ADSL) telephone wires.) Other cable modem vendors such as General Instrument, Intel, LANcity, Zenith, and Motorola also have proposed cable modems with speeds ranging up to 30 megabits per second downstream and 10 megabits upstream.

Focusing on two-way Internet access, Figure 7.6 illustrates the hybrid fiber–coaxial cable (HFC) configuration now regarded as the state of the art in cable system design. Cable modems will use shared upstream and downstream bandwidth. The capacity of a cable system to supply bandwidth for this purpose depends on how many channels are allocated to it and on the ratio of nodes to subscribers. As demand increases, cable systems can allocate more channels, reduce the number of subscribers sharing each fiber node, or upgrade the system as a whole.

In order to install cable modems to serve subscribers in HFC systems, it is necessary to make provisions for one or more channels (frequency bands) for the data traveling in each direction. Amplifiers must be installed in the upstream direction, and the system must be fine tuned against various technical problems such as increased vulnerability to interference. Under the standard configuration for modern cable systems, the frequency band from 5 to 40 megahertz is available for upstream data flow; this band, however, is exceptionally "noisy." Figure 7.7 illustrates proposed and actual uses of the capacity of a modern cable system, including digital and analog TV channels, support for mobile telephone service (PCN), and upstream and downstream data flows associated with Internet use.

Cox Communications, a cable MSO, has been among the leaders in upgrading systems to support two-way cable modems. Cox introduced cable modem–based Internet access service in Orange County, California, in 1996, and extended it to San Diego and other cities in 1997. The sophisticated Orange County system, which not coincidentally competes with Pacific Bell's nascent digital wireless cable system, also offers local telephone service.

Most of the data flow in a domestic Internet connection is downstream to the subscriber. Data flows in the reverse direction are much

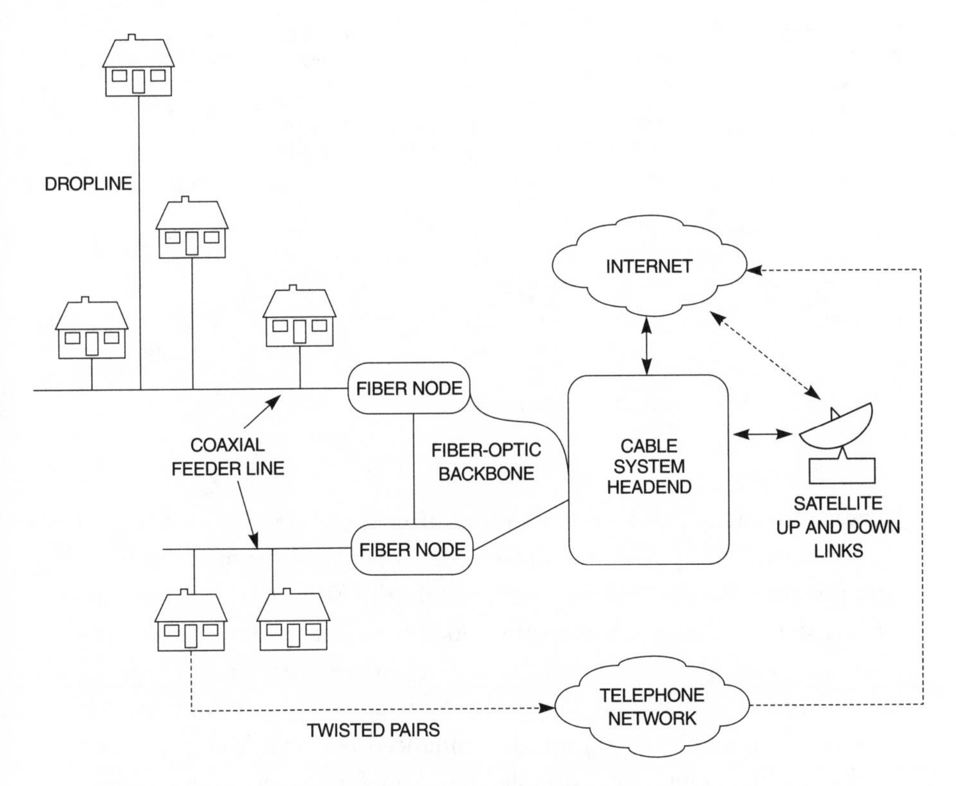

Figure 7.6 A hybrid fiber-coaxial (HFC) cable system

smaller. Therefore, the cable operator can cut much of the cost of offering Internet service (including, in many cases, the need for an HFC upgrade) by offering only one-way service: downstream to the subscriber. The subscriber then must find another means to send data out into the Internet. At present the only realistic choice is the telephone.

Most of the cable modem designs under development or designated for early deployment are for one-way service. Although one-way service may reduce the cable operator's investment and risk, it increases the subscriber's cost: both an analog modem and a telephone line are required as well as a cable modem and Internet access service. As of early 1998, fewer than 100,000 United States households were served by two-way cable modems.

One-way cable modems (Figure 7.8) seem to be the technology of

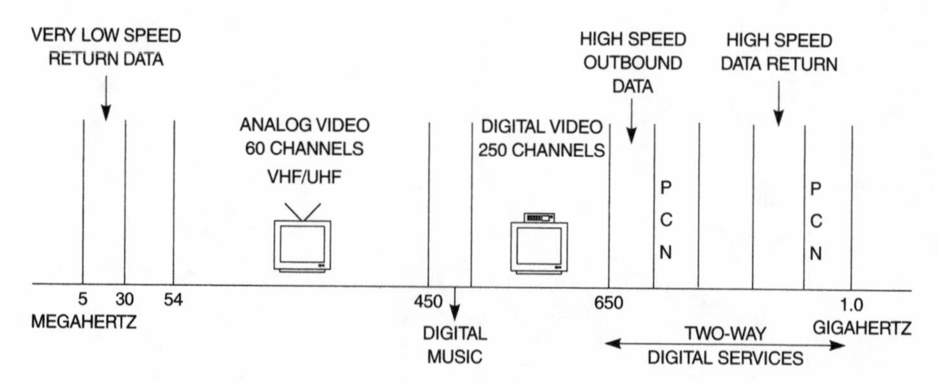

Figure 7.7 Future uses of cable bandwidth

choice for many cable operators. But after spending around $50,000 for headend equipment and around $500 per user for modems, they may receive about $30 per user per month. Although one-way modems may be less risky for cable operators, few experts expect the service to achieve high penetration rates. Moreover, many challenge the assumption that data flow will be unbalanced or "asymmetric." If the Internet is to be used for video-enhanced personal Web pages and video conferencing, for example, the data flow may be nearly balanced. In this case the present model for cable modem–based Internet access will fail, and it will fail essentially because cable operators will have held on to the "broadcast model" of the business. To succeed, Internet access service must be two-way broadband interactive service—in effect, a "telephone model."

If cable operators decide to upgrade their systems to offer two-way service, the issue of asymmetry remains. As noted earlier, the frequency band from 5 to 40 megahertz is available for upstream data flow. A bandwidth of 35 megahertz, even though shared by many users, is a lot for ordinary data such as "mouse clicks" on Web pages. But it is not a lot for two-way broadband service. A few subscribers sending out 50-megabit-per-second video bitstreams could easily overwhelm this capacity.

HFC cable systems have fiber-optic backbones and coaxial cable running from neighborhood nodes to individual homes in a tree-and-branch structure. The fiber backbone can readily handle high-bandwidth upstream flows, and the drop line to the subscriber's house can

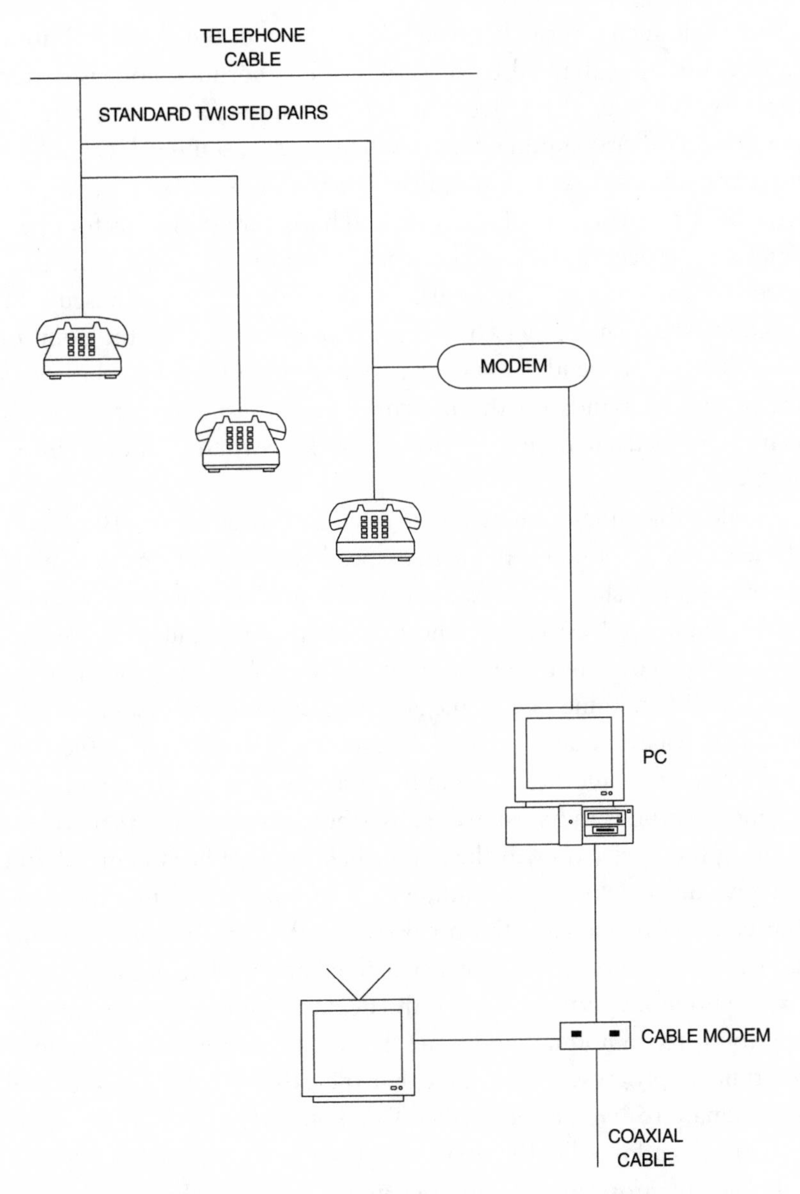

Figure 7.8 One-way cable modems require use of the telephone system

also handle such service, because it is not shared. But the distribution cable connecting all the drop lines to the neighborhood node is shared capacity. When used for two-way Internet access, a cable system resembles an office computer network. The present standard technology for two-way data service on such a network is called "Ethernet," which is a particular set of software and hardware standards for operating a local packet-switched network. Most of the technologies proposed for cable modems and the cable system are Ethernet-based. The speed at which such a network appears to operate, from the point of view of a user, depends on its capacity and on how many other users are putting demands on the network. The transmission capacity is shared, not dedicated, just as with the packet-switched Internet backbones.

While cable modems may be advertised at speeds up to 10 or even 30 megabits per second, the actual speed will depend on how many other users are sharing the same capacity at the same time. A cable system configured with a tree-and-branch structure and with present-day capacity could not support hundreds of cable modems in simultaneous use at anything like 10 megabits per second. Increases in capacity (or increased node density) will certainly be required if cable is to be the primary vehicle for interactive video delivery to the home.

One solution is to have fewer subscribers per node: in other words, to bring the fiber closer to the subscriber. Because both demand and exact means of delivery are unknown, however, this is quite expensive. The effect is to increase the break-even subscriber revenue for this service, a service yet to be precisely defined and for which the demand may be no more permanent than the demand for CB radios. Another solution is to do nothing and hope that data compression technology will quickly progress to the point that the existing cable facilities will be adequate to handle the traffic. This is the solution that the cable industry has adopted in the past several years.

Cable operators' main line of business is delivering broadcast-quality video signals. When adopting a cable modem approach, they must keep in mind how it will affect this business. Problems lie in simultaneously choosing the correct data compression and the correct transmission technology for the next generation of cable video services; the decisions are not independent. The issue of high-definition or highly compressed digital television is discussed more fully in the broadcasting context in Chapter 15. Although the FCC has imposed a compression standard on the broadcasting industry, that standard need not be adopted by the cable industry. It is this technology that will, perhaps, turn all those remaining 40-channel and 80-channel systems into 550-megahertz-and-up 500-channel systems. (According to CableLabs, a state-of-the-art gigahertz digital HFC cable system has a capacity of 160 6-megahertz channels, each of which can carry four compressed video signals of current broadcast quality. The National Cable Television Association has touted compression ratios of "up to" 12:1.)

The cable modem problem and the cable video capacity problem are related. Since the answer to the latter has been repeatedly delayed, so likely will be the former. Moreover, the solution to these problems that is correct for any given cable operator depends on what other cable operators decide to do. For various reasons, including economies of scale in the production of equipment and economies of standardization in program transmission, there are "network externalities" in these decisions.

The Race and the Contestants

The cable industry is in a race to provide the cheapest broadband interactive transmission capacity to households. Cable may not win the race even if it settles on what turns out to be a good choice for its own technology. There is every chance that telephone companies, or satellite companies, or wireless cable companies may cross the finish line first.

The cable industry, however, has two advantages in the contest. First, it is free to select and change its technical standards. Broadcast media standards can be changed only by the FCC. Cable's flexibility is a boon in a world of changing technology. Second, cable is the only existing hard-wired, broadband communication infrastructure that is already universally available in the United States. Whatever adaptations it must make will be incremental, rather than from the ground up.

The cable industry also has two fundamental handicaps in the race. First, cable modems and associated set-top digital converters may not be the most cost-effective way to provide interactive broadband services to consumers. Telephone companies employing digital technology can build fiber to nodes (or even "to the curb") and offer fully switched service. Other technologies such as wireless cable are also contenders. Second, even if cable modems are the technology of choice, the cable industry's geographically fragmented structure makes it very difficult to organize a coherent marketing effort in competition with telephone, satellite, or gigahertz-band wireless cable services. Regional consolidation of MSOs may make a lot of sense, both for the industry and for consumers, but might face stiff opposition in Washington.

Conclusion

The cable television industry came into existence to satisfy consumer demand for programming that could not be satisfied by conventional broadcasting because of FCC channel limits. Cable is probably a less efficient way to provide programming than conventional broadcasting with equivalent channel capacity. If so, the cable industry would not exist but for restrictive spectrum policies.

The widespread availability of cable came about not only because of unmet viewer demand for programming but also because the advent of C-band communication satellites greatly reduced the cost of national distribution of programming to cable systems.

Cable operators traditionally used coaxial cable all the way from their headends to the consumer's television set. Modern systems use fiber-optic cable at least for the main trunks. These have great channel capacity. To bring fiber all the way to the home, however, would be extremely expensive given the lack of assured consumer demand for interactive video services. Hybrid fiber-coaxial systems are therefore being constructed, with fiber running to neighborhood nodes, and coaxial cable from nodes to subscribers.

The cable industry has had a rocky relationship with government regulators, being alternately promoted and condemned, regulated and deregulated. The cable industry was most recently reregulated in 1992 amid accusations of price gouging and monopoly, despite the fact that over the preceding period of deregulation output had greatly expanded and unit prices had fallen. One element of the ambivalent cable industry–government relationship rests on the unrealistic expectations created a generation ago by those who saw cable as the "information highway." Analogous expectations with respect to the Internet may be forming now.

Cable's ability to provide interactive video services, Internet access, or telephone service hangs in the balance today. Few cable systems can offer any of these services. Even so-called cable modems that permit one-way Internet access are not widely available. Systems that can offer two-way service remain chiefly experimental. Upgrading cable systems to provide such services is extraordinarily expensive. Cable companies do not yet know which technology is best for this purpose, in part because they do not know the extent of consumer demand.

CHAPTER EIGHT

EARLY DIRECT BROADCAST SATELLITES

There are two generic methods to transmit electronic signals to the home: wires (telephone or cable) or wireless (radio waves). Within the wireless category transmissions can occur by terrestrial broadcast (which includes ordinary television and radio signals, but also many newer technologies) or by satellite broadcast. This chapter is concerned with satellites that broadcast directly or indirectly to the home by means of C-band (3.4 to 7.0 gigahertz) geosynchronous communication satellites, or GEOs. This is an analog technology. Digital communication satellites and other digital media are discussed in Part IV.

The Clarke Orbit

Geosynchronous or geostationary satellites travel at exactly the same speed as the rotation of the Earth and therefore appear to be tethered permanently in one spot above the equator. There is only one altitude that permits this: the satellites must be about 22,300 miles above the Earth (see Figure 8.1). A geosynchronous orbit is sometimes called a "Clarke" orbit after Arthur C. Clarke, the science fiction writer who first conceived of it in the 1940s.

Because the satellites are in one position, receiving antennas on Earth can be relatively inexpensive. (They don't need steering devices to track the satellite across the sky.) This reduces the overall cost of a satellite transmission system (satellites plus ground stations). One drawback of a geosynchronous satellite is its altitude: because it is so high above the Earth (much higher, for example, than the space shuttle missions), it takes about half a second for a radio signal to make the round trip, even at the speed of light. This interval is perceptible and, to some, annoying, in telephone conversations. It can also be a problem for certain kinds of data transmission, especially real-time interactive video. Imagine playing an interactive arcade game on the Internet when each move requires a half-second lag.

Transmissions are beamed ("uplinked") to a GEO from one or more Earth stations. On board the satellite the signals are amplified, shifted in frequency, and broadcast to anyone capable of receiving them within the "footprint(s)" of the satellite's downlink antenna or antennas (see Figure 8.2). In a sense, a GEO satellite is like a microwave relay tower, picking up signals from adjacent towers and retransmitting them. A GEO broadcast satellite may have one or more "global" beams that reach the entire portion of the globe (about one-third) visible to it, or the northern or southern hemisphere, and it may have one or more "spot beams" that reach a more circumscribed area, such as the northeastern United States.

In the case of spot beams, the frequency band available to the satellite may be reused. For example, a satellite may be able to broadcast 200 channels on a global beam to the entire continental United States, or it may be able, using the same frequencies, to broadcast 200 channels to the U.S. East Coast and 200 different channels to the U.S. West Coast. In other words, frequencies can be reused just as they are in cellular telephone systems. The frequencies can be reused because the beams do not permit the signals to interfere with each other. In theory, it would be possible to beam 200 different channels or an equivalent amount of data to each household. Moving

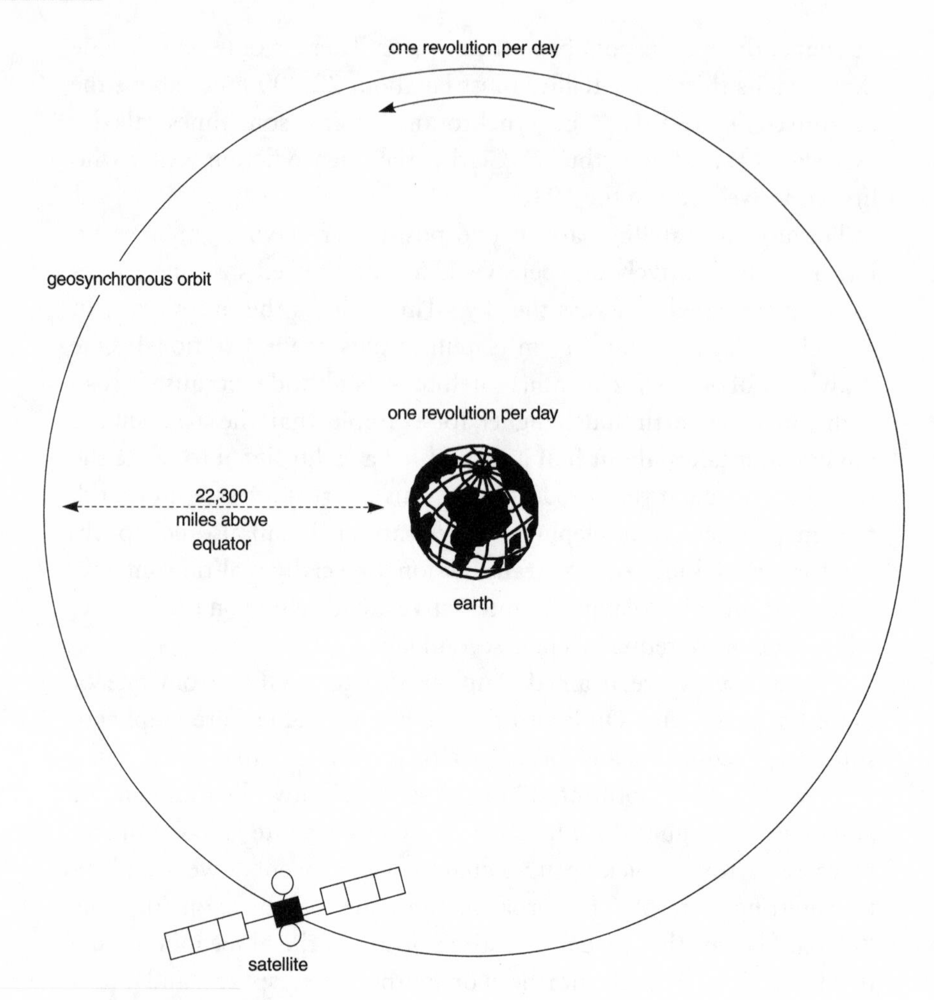

Figure 8.1 The Clarke orbit

from global beams to spot beams for a given channel capacity, however, increases the satellite launch weight, power requirements, and station-keeping challenge. These are all costly considerations. As a result, it is not practical today for a single satellite to have a spot beam aimed at every home or even every city. Multiple satellites working together, however, could achieve coverage of dozens of relatively small regions.

The paddle-like structures on the satellites in Figure 8.2 are solar

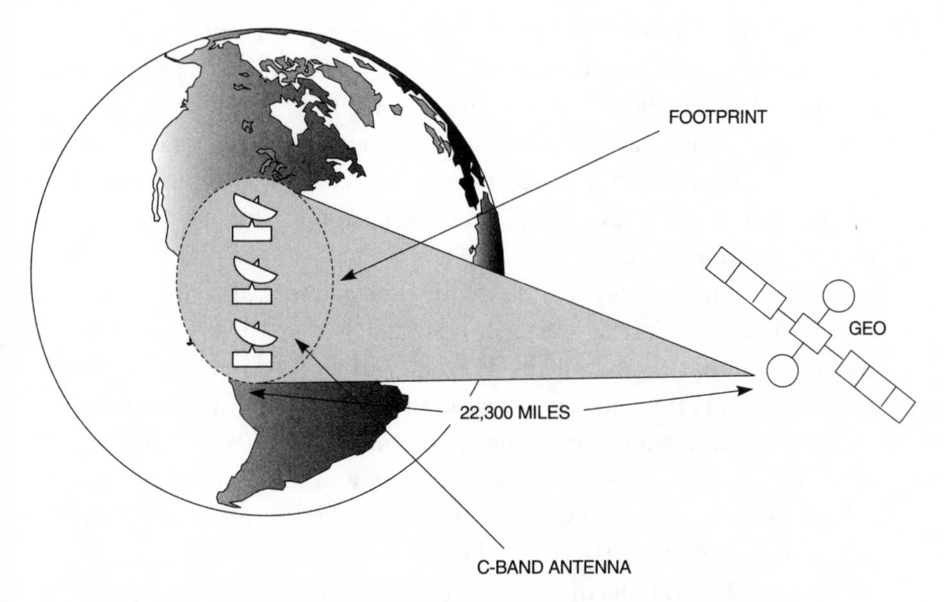

Figure 8.2 Communication satellite footprint and receiving antennae

cells used to power the satellite. The round appendages are antennas shaped to illuminate specific geographic areas.

The United States is a member of an international body (the International Telecommunications Union, or ITU) that negotiates uses of particular frequency bands and orbital slots (satellite parking places and the separations between them). As a result, the frequency bands available for use by U.S. firms and the number of orbital slots are fixed by treaty. Both are constraints. If it were not for the limits on bandwidth, U.S. direct broadcast satellites would have more channels. And, if it were not for the constraint on slots, more satellites would serve the United States. (Given the separation standards, there is a limited number of orbital slots from which a satellite can "see" all of the United States. Some of these slots are allocated to Canada and Mexico, among other countries.)

The first generation of GEO television broadcast satellites used the frequencies from 3.4 to 7.0 gigahertz, the so-called C-band. These satellites were intended to transmit video signals from broadcast and cable networks to local TV stations and cable systems, which then rebroadcast them to consumers. The use of satellites as trunks to connect local distributors to points of content origin, rather than as direct links to the home, is a feature of many current and proposed technologies for video and telephone communications.

C-band satellites are relatively low powered, and that, taken together with the frequency band, dictates a receiving dish from 7 to 30 feet in diameter. Dishes of such size are unsuitable for use by most suburban and urban households.

The first commercial C-band satellite designed for video broadcasts within the United States, Westar I, was launched in 1974. Owned by Western Union, it was built by Hughes and could carry twelve TV channels. It was this satellite, along with HBO, that began to revolutionize the cable television industry. During the 1970s, C-band became accepted as the most cost-effective way to distribute video signals to TV stations and cable systems; previously this service had been accomplished by AT&T terrestrial microwave facilities, and by sending videotapes and films through the postal service or by messenger.

Unexpectedly, as the price of C-band receiving equipment and dishes declined (from about $10,000 in 1980 to around $3,000 by the end of the decade), and as the number of satellites and channels increased, ordinary consumers began to acquire such systems. These "home satellite viewer" consumers were chiefly rural, too far away from urban areas to be able to get either broadcast or cable television. Even though production volume increased, the equipment remained expensive—typically well over $2,000—and the antenna was so big (usually 8 or 9 feet in diameter) that only rural backyards could readily accommodate it.

As the number of consumers equipped with C-band antennas increased, approaching an estimated one million by 1985, it became apparent to television and cable networks that these consumers were "pirating" the networks' copyrighted video signals. As a result, C-band signals were encrypted, and consumers equipped with C-band antennas were required to purchase special decoder units and to pay a monthly fee to receive most of the networks. The encryption technology was supplied by General Instrument's "Digicipher" devices. General Instrument maintains a control center that sends coded signals via satellite to set-top boxes in the home, and these signals unlock the channels for which customers have paid. There has been a substantial cottage industry pirating these devices. As of January 1997, there were

2.3 million C-band households in the United States. This number is now beginning to decline because of the superiority of the new Ku-band digital services.

As of late 1996 there were twenty-three C-band GEOs serving the United States with a collective bandwidth of 2,600 megahertz, or the equivalent of 433 6-megahertz TV channels at each of two polarizations, for a total capacity of 866 channels before any compression. These satellites are not used exclusively for broadcast television, however; they also serve communication needs such as point-of-sale data networks (connecting all K-Mart stores to the home office, for example) and as transmission paths for mobile television news and sports crews sending video signals to TV stations or networks for editing and rebroadcast.

In addition, C-band satellites are used to provide multichannel video service to hotels, other commercial establishments, such as sports bars, and local TV distributors who compete with cable television. There are two categories of these "private" or unfranchised cable systems: those serving apartment buildings and subdivisions (called SMATV, or satellite master antenna television), and "wireless" cable systems (one of which is called MMDS, or multichannel multipoint distribution system). Both of these cable rivals use analog technology and rely on C-band satellites.

Local Distribution Hanging on a Satellite: Private Cable

Cable television systems are coterminous with local political subdivisions, each of which has the power to grant franchises. These franchises in practice have been exclusive. They do not extend to private property, however, because the city's franchise is a right to use the public streets. The most profitable areas for a cable company are high-density population concentrations, especially in new construction or construction designed to be easily wired. These are also the most profitable areas for anyone supplying multichannel video services. Therefore, apartment building owners and subdivision developers sometimes install private "satellite master antenna television" systems (SMATV).

SMATV systems operate under immunity from local franchise regulation and from federal rate regulation. In reality, they are no more or less monopolistic than cable systems. The FCC figures that apartment owners (who must compete with one another for tenants) will compete away in lower monthly rents whatever monopoly profits they may make on the SMATV service. In practice, many apartment owners bundle SMATV with other building services, making no separate charge to tenants.

Subscriptions to the various C-band cable networks can be purchased by an individual directly or from one of two major packagers, one owned by TCI and the other by Viacom. SMATV systems obtain television signals from a C-band antenna, having made arrangements with one or more cable networks or packagers to subscribe to popular program services. Some SMATV systems, however, obtain their video feeds from MMDS operators. Others simply make a deal with the local cable system for a bulk rate discount.

A Taste of the Future: Analog Wireless Cable (MMDS)

Multichannel multipoint distribution systems also are made possible by C-band satellites. MMDS systems offer almost precisely the same services as a cable TV system, except that they deliver the service using microwave band radio signals in the 2.4-gigahertz range. Users of this frequency band were limited, when the service was first permitted by the FCC, to around six standard TV signals. But the FCC later increased the bandwidth available for MMDS service, and of course MMDS operators are free to use compression technologies as well. Many analog MMDS systems now offer more than thirty channels. There is a single MMDS operator in 493 U.S. cities, but the service has never been extremely successful; its major focus has been on feeding video to SMATV sites rather than to individual households.

One reason for the lack of success of MMDS relative to cable television is limited channel capacity. Although it has increased, the channel capacity lags behind that of cable and direct broadcast satellites. Moreover, the frequency band available to MMDS operators is more limited than the bands available to other wireless service providers. Another handicap has been the need to install special antennas on subscriber rooftops, along with electronic equipment to decode the signal. The cost of the special equipment and installation offsets most or all of the advantage of not having to run wires down the streets.

Finally, the microwave bands are subject to degradation due to rain and other factors, causing reception problems.

Telephone companies have flirted with MMDS as a means to provide local video services in competition with cable television systems. The flirtation cooled when it became clear that cable systems were not going to enter the telephone business anytime soon. However, PacBell has converted several MMDS systems in California to digital transmission at a cost of more than $150 million, greatly increasing channel capacity, and it appears ready to compete vigorously with cable in those areas.

From Sputnik to Open Skies

Communication satellites were an early focus of the cold war space race between the United States and the Soviet Union. It was the launch of Sputnik that galvanized the United States into reaching for the moon. Although communication satellites have both military and civilian uses, a major goal of the U.S. space program was to develop civilian applications. Civilians owned and operated communication satellites used for nonmilitary purposes, even though government-owned launch facilities were needed. AT&T did much of the initial development work on communication satellites, and one early issue was whether to permit AT&T to obtain or retain a monopoly in this area. The first civilian uses of communication satellites were for international telephone calls, which required connecting to foreign telephone systems, almost all of which were then owned by national governments. As a gesture of goodwill, the United States decided that, rather than have a private U.S. monopoly deal with other nations concerning these arrangements, these fruits of space race research should be shared with the entire free world.

Accordingly, in the early 1960s, the United States organized the formation of an international consortium (INTELSAT) that would own all international communication satellites. The U.S. "representative" to INTELSAT was a quasi-private corporation, COMSAT. It was given a monopoly (which it still retains) on the right to provide international satellite telephone service to and from the United States. INTELSAT voting rights and investment shares were tied to usage, and for a long time COMSAT had a majority interest in INTELSAT and was its "manager," making it the free world communication satellite monopolist. (In the early years, AT&T in turn owned a large part of COMSAT's common stock, but this was divested around 1970.) INTELSAT provided satellite-based telephone service to the free world in "competition" with INTERSPUTNIK, which supplied such service to the Soviet Union and its allies.

U.S. manufacturers of communication satellites in the 1960s had only one customer for international service (INTELSAT), and only a handful for domestic service (AT&T and telegraph companies, such as Western Union). Research by these manufacturers resulted in increased satellite capacity, however, and by 1970 communication satellites could relay TV signals rather than just telephone calls and telegrams. The question then arose: who should be permitted to own and operate domestic communication satellites (DOMSATs)? AT&T argued that DOMSATs were an integral part of its natural monopoly telephone system. The satellite manufacturers argued that it would be better for them to sell into a competitive market.

President Lyndon Johnson convened a special White House panel to consider this and other communication policy issues. His successor, Richard Nixon, acted on one of the panel's recommendations and appointed what would today be called a communication "czar" as part of the White House staff. Not since Herbert Hoover had the executive branch been so deeply interested in broadcasting. This czar, Clay T. Whitehead, succeeded in formulating and implementing a domestic "open skies" policy, under which anyone would be able to own and operate a domestic communication satellite. The FCC somewhat grudgingly adopted this policy in 1972, and later it was codified. Accordingly, for the first time ever, the United States had an opportunity for free competition in long-distance broadband communication. The beneficiaries of this policy were cable systems and broadcast networks (who would no longer need to rely solely on AT&T's expensive television microwave relay service) and satellite manufacturers (who would not have to deal with a monopsonist).

Making Money with C-band Satellites

C-band television relay satellites were developed initially by Hughes, building on a decade of experience in constructing communication satellites for telephone companies, for INTELSAT, and for the military. The economic viability of C-band depended on the cost of satellite transmission of TV signals and the price charged by AT&T for its terrestrial microwave television relay service.

The standard business plan in those early days was for a company such as Western Union or AT&T to purchase a satellite (and usually two spares, one to be launched into orbit in case the first one failed and the other a back-up on the ground), to pay NASA for a launch, and then to offer service on the satellite to customers on a common carrier (tariffed) basis. This required considerable up-front capital. Today satellites and launch services together can easily run to $200 million, and the cost in real terms was even higher in the 1970s, before France, China, Russia, and others began to compete with NASA in the launch business.

Because satellites were so expensive and because no one television network customer had need for an entire satellite, it appeared that some entity would have to serve as middleman or broker, buying satellites and selling satellite services. Hughes did not wish to be in the business of selling communication services, perhaps because it would face greater risk selling its satellites in this way. However, in the mid-1970s a Hughes employee named Clay T. Whitehead (the former czar) hit on the idea of selling satellite "condominiums." Communication satellites are made of basic units called transponders, which receive signals on one frequency band and send them out on another. In the early satellites, one transponder corresponded to one TV channel; in later satellites two TV channels were put on each transponder by means of a shift in the polarization of the signals. Whitehead sold condominium interests in individual transponders to companies like CBS and HBO. Typically, the transponders were sold out prior to launch. In this way, Hughes increased the overall demand for the satellites that it and others manufactured while substantially reducing its own commercial risks.

There followed a great boom in the C-band satellite business, in concert with an explosion of new cable networks, and the growth of cable penetration in urban areas. Cable systems, cable networks, and satellites were all heavily dependent on one another, and yet each developed as an independent industry. That is, the very obvious "network externalities" were realized in spite of the fact that there was relatively little vertical integration between the three industries. True, some multiple system operators (MSOs) such as TCI owned minority interests in certain cable networks. And of course a number of cable networks "owned" or leased satellite transponders. But these ownership interests had little to do with the growth of the three industries. This is a lesson worth remembering when, in discussing the new digital media, we consider the problems faced by new communication technologies with similar chicken-and-egg issues.

Conclusion

Geosynchronous communication satellites hover 22,300 miles above a fixed point on the Earth's equator, permitting antennas on Earth to maintain a fixed orientation. GEO satellites operate in the C-band or microwave frequencies at 3–7 gigahertz. Such satellites are so successful and popular that there is a limited number of orbital slots and frequencies available for their use.

C-band GEO satellites have been particularly important as a means of distributing television signals to cable systems and to broadcast stations. They are somewhat less satisfactory for two-way services such as telephone calls because it takes a perceptible time, even at the speed of light, for the signals to travel to and from the satellite.

In addition to supplying cheap program distribution service to cable systems and broadcasters, C-band satellites also supply programming to private cable systems servicing apartment complexes and institutions, and to MMDS or wireless cable systems.

The success of C-band satellites has rested on government "open skies" policies that have permitted competitive private development rather than regulated monopoly. Despite chicken-and-egg problems associated with the uncertainty of demand for satellite services, extensive vertical integration has not taken place.

PART THREE

THE DIGITAL COMMUNICATION WORLD

CHAPTER NINE

UNDERSTANDING DIGITAL MEDIA

From the point of view of digital communication technology, "information" is a "bitstream" of zeros and ones, just like an old-fashioned dit/dah telegraph code. Digital technology works the same way regardless of content. Anything that can be digitized can be transmitted on a digital medium to one and all. And virtually any image can be digitized, including books, newspapers, paintings, movies, TV programs, music, personal conversations, speeches, political cartoons, brainwaves, and three-dimensional objects. Further, almost any current electronic communication medium is or can be made into a digital medium, including telephone systems, television broadcasting systems, cable television systems, and geosynchronous communication satellites. It is easy to see why convergence is a focal concept: everything seems headed in the digital direction very rapidly. But this may be an illusion.

Communication is the transfer of information. We pay to transfer information because it is useful or entertaining in form or substance. For example, when Amazon, the Internet on-line bookstore, sends me a book through the mail, there is a communication of information (the content of the book). This is no less a form of communication than a telephone call or a TV broadcast, and there is no logical basis for thinking differently about what has happened than if the content were mailed in the form of a compact disk or floppy disks; transmitted in the form of a standard text file attached to an e-mail message; transmitted over the World Wide Web and displayed in a standard format on a computer screen; read during a broadcast television program; or acted out in a film played in a movie theater. Understanding that all of these media are at the disposal of any communicator, to be selected on the basis of both economic and aesthetic considerations, helps to put into proper perspective many of the issues that arise in the new electronic media.

Information, however, is a concept more subtle than the "content of the book." Data are information only if they tell us something new or change our perception of reality. Suppose I already know everything in the book except one important fact that appears on page 567. Then the fact on page 567 is the only information that has been transmitted to me though the medium of the postal service and the book. The remaining content of the transmission is redundant baggage.

Information Economics: Just the Facts, Ma'am, Just the Facts

Why would anyone want to send or receive redundant baggage through the mail? Why not just send or request the particular fact on page 567? There are several possible reasons. First, it may cost more to pay someone to find page 567, copy it, and fax or mail it to me than to order the book. Economies of scale in packaging data may make it cheaper per "fact" to produce and transmit facts bundled together. Second, I may not know what page the fact is on, or even whether it appears in this book. My ordering the book may be part of a search strategy. Third, I may wish to refer to this fact in the future, and a book on the shelf is a reliable place to store it. Fourth, the publisher of the book for marketing reasons may be unwilling to "sell" (permit third-party copying of) pages from the book. Finally, page 567 (or the fact found there) may be an integral part of an aesthetic whole that lends added meaning and value to the fact itself.

In spite of all these reasons why I might not request or send page 567 by itself, it often is more efficient to send page 567 alone, and modern communication technology is largely concerned with finding ways to do that. We will return to this issue below in the context of "data compression."

The publisher's marketing considerations are worth a closer look. Viewed from an economic perspective, the book transaction involves the sale by the publisher of a fact to a particular reader. There are two considerations from the publisher's perspective. First, what is the cheapest way to transmit the fact? In other words, what *medium* will have the lowest per unit transmission cost? Second, in what *form* will the fact be most useful to the recipient? Put differently, how can the publisher package the fact in a way that makes the recipient willing to pay the most for it? In practice, the publisher has imperfect knowledge of the recipient's preferences, so marketing strategy involves some groping in the dark and a lot of bundling. Bundling sometimes works in this context because it permits publishers to extract maximum revenue from consumers even though the exact valuation placed on each fact by each consumer is unknown. Moreover, in general, these two considerations—medium and form—are not independent. The medium conditions the form of the message, and both affect the cost of the transmission; a lower-cost transmission technology may reduce the value derived from the form of the data by more than the cost savings in transmission.

Here is a different example. A number of years ago (long before the *New York Times* finally introduced color in 1997) it became possible to include color graphics of reasonable quality anywhere in a daily newspaper. Publishers had to decide whether a change in the *form* of the newspaper (the inclusion of color graphics) would increase the willingness-to-pay of subscribers and advertisers, and if so whether the increased revenues would cover the cost of the new press equipment. There were several types of press equipment available, of course, and the more expensive equipment could produce higher-quality color images. Which new press should a publisher buy, and when?

In the 1920s and 1930s, over-the-air television broadcasters faced similar choices. The technical quality of the image that appears on the

TV screen depends on the frequency, power, and bandwidth of the broadcast signal, the characteristics of the sending and receiving antennae, the distance from the transmitter to the receiving antenna, and the quality of the receiving equipment (the TV set). Image quality depends on the technical standard employed to generate and reproduce the video image (frames per second, lines per frame). Image quality also depends on the content of the broadcast. Sports events generate more "information" per second than love scenes, and so a channel of given bandwidth will be able to reproduce the latter more reliably than the former. Like newspaper publishers choosing a color press, television pioneers had to decide how to balance many considerations.

The Meaning of "Bandwidth"

A closer look at the decisions faced by the television pioneers can help us understand bandwidth (in its popular sense of channel capacity)—the current topic in discussions of new media. First, consider the problem of choosing a transmission medium for television pictures. In the 1930s there were two potential ways to transmit video signals: over-the-air broadcasting and coaxial cable (what is today called cable television). Obviously it would be quite expensive to run cables from each local TV studio to everyone's home, even if all the stations used the same cable. There was no such cable in existence, and no institution to organize its construction; in any event, construction would take many years. (Although cable television came into existence at virtually the same time as over-the-air television broadcasting, it was limited at first to rural areas and to retransmission of over-the-air signals.) In contrast to cable, the airwaves were available and free. Nevertheless, the airwaves had some disadvantages compared with cable. For example, for a given bandwidth the quality of reception is higher on cable because it is less subject to outside interference. Balancing these and many other such considerations, the television pioneers chose over-the-air broadcasting as their medium, along with a particular method of analog transmission (known as NTSC after the National Television Standards Committee that adopted it), a method entirely incompatible with television standards elsewhere in the world. This method of transmission remains in place with some modifications to this day.

Now consider the problem of choosing a bandwidth for a TV broadcast. Bandwidth is roughly analogous to the diameter of the pipe in a plumbing system. Other things equal, the greater the bandwidth the more data can be transmitted. For a given data transfer rate, greater bandwidth increases image reception quality and therefore reduces the complexity and cost of receiving equipment.

The pipe analogy is a useful way to understand bandwidth and its significance. Imagine that two people stand at opposite ends of a 50-foot-long, ½-inch-diameter copper pipe. They attempt to communicate by shouting through the pipe. The communications must be one way, because with such a small pipe one's ear and mouth cannot both address the pipe simultaneously. The transmissions will be distorted and difficult to understand, and messages may have to be repeated many times. Now suppose the 50-foot-long pipe is 8 feet in diameter—a culvert. Our communicators can stand upright in it and understand each other with far greater ease. Another use for a big pipe is to transmit more data, rather than less data more clearly. Our communicators in the 8-foot-diameter pipe could transmit a "multimedia" presentation by holding up colorful posters or even projecting a slide show from one end of the pipe to the other, while shouting as before.

Bandwidth in this example is used to mean the "capacity of the pipe" in units of data per unit of time. The bandwidth of a daily newspaper printing press is measured by its maximum number of pages (or equivalent column-inches) per issue-cycle of the newspaper. In electronic communications, as noted above, bandwidth has the narrow technical meaning of the difference between the highest frequency and the lowest frequency on a continuous frequency range that the medium employs. Frequency used to be measured in cycles per second, now called hertz. Imagine a pure signal or tone, either broadcast or on a cable, that vibrates at exactly 1,000,000 hertz (1,000 kilohertz or 1 megahertz). The same source may also broadcast at another tone, say 1,000,010 hertz (1,000.01 kilohertz or 1.00001 megahertz). Such a source can broadcast at *every* frequency between these two signals—that is, on the whole range from 1.00000 to 1.00001 megahertz. This continuous range of frequencies (10 hertz wide in the example) is called the "bandwidth" of the signal.

A bandwidth of 10 hertz is extremely narrow. An FM radio signal

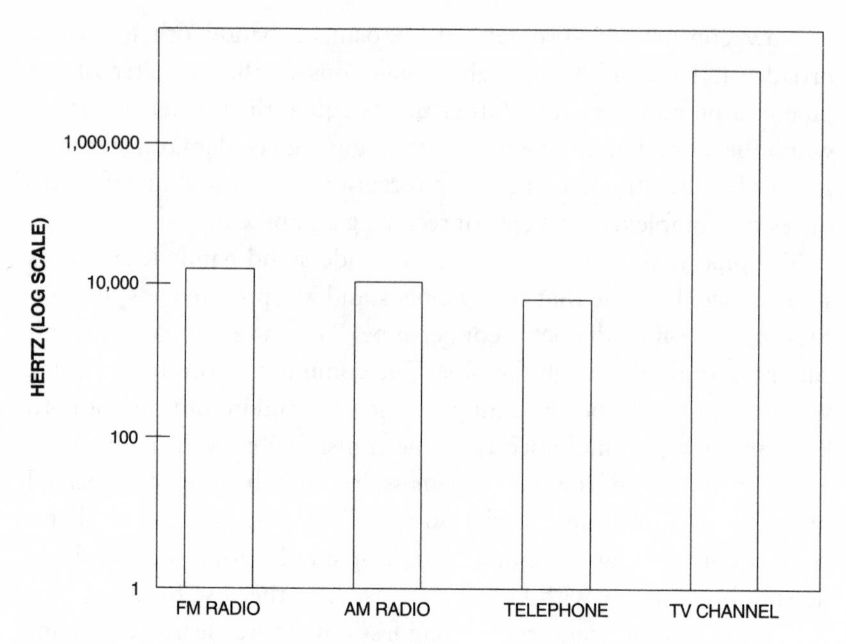

Figure 9.1 Bandwidths of different media

has a bandwidth of 15,000 hertz, and a television channel has a bandwidth of 6,000,000 hertz (see Figure 9.1). But in another sense, 10 hertz is very broad, because there is an infinite number of specific frequencies in a 10-hertz band. No matter how close two specific frequencies may be (say, 1,000,000 hertz and 1,000,000.9999 hertz), there is always another frequency between the two, such as 1,000,000.99995. (Indeed, there is an infinity of frequencies between the two.) Broader frequency bands are employed, not because data cannot be transmitted on narrow bands, but because usually it is cheaper to use broader bands for a given amount of data. Audio data that can be heard clearly by the unaided ear in a 12-foot-diameter culvert would require expensive and sophisticated electronic equipment to be detected and interpreted in a ½-inch-diameter pipe.

When people talk about "broadband" communications, they generally mean the practice of sending a large amount of data, such as one or more video signals, down a large pipe. But it is important to keep in mind that the "large pipe" is a convenience, a matter of economics, and not the inevitable requirement of transmitting a large amount of data. At increased cost, the same data could always be sent on a narrower band. Naturally at some point the limits of current technology would be reached, although as we shall see, technology responds not only to scientific discoveries but to economic incentives as well.

Bandwidth is expensive and sometimes not available at any price. Eight-foot-diameter culverts cost much more than ½-inch-diameter copper pipes, per unit length. In a cable, capacity is mostly determined by the equipment at the ends of the cable, and by the amplifiers along the way. In any over-the-air broadcast, the bandwidth and capacity are both determined by the FCC. In a cable, more capacity requires more expensive equipment. In nature, or in the market, more capacity can always be purchased, at some price. By contrast, in broadcasting, more capacity can be purchased (if at all) only through lengthy, tedious, and expensive dealings with the FCC and with the political representatives of rival claimants for the bandwidth. Tellingly, when the government auctioned spectrum for new digital cell phone services, one of the few restrictions was that the frequencies not be used for broadcasting.

Because bandwidth is expensive, it is not sensible to use an arbitrarily wide band to send any given information, even though that would reduce the cost of the equipment needed at both ends. There is, in other words, a tradeoff between bandwidth cost and equipment cost (see Figure 9.2) that is really a special case of the general tradeoff between processing, bandwidth, and storage illustrated in Figure 2.1. If a signal is spread out across a broader band, the likelihood that the receiver will be unable to detect it (on account of interference, say) goes down, and so the receiver can be built in a less discriminating and thus less expensive way. If the same amount of data is sent on a very narrow band, sending and receiving equipment must be much more elaborate in order to distinguish the signal in question from ambient interference and other degradation. To abandon the pipe analogy for a moment, the tradeoff is almost the same as the difference between reading proverbs in a book of proverbs and reading proverbs through a scanning electron microscope because someone, in order to save space in the delivery truck, has gone to the trouble of spelling them out with individual atoms on a tiny piece of silicon.

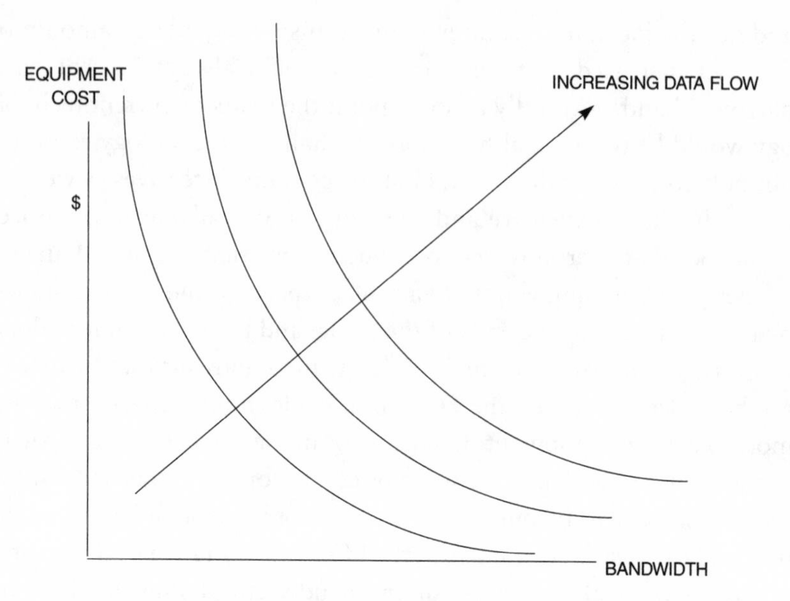

Figure 9.2 Bandwidth-equipment cost tradeoff for given data transfer rate

Shouting over the Din

If we are shouting at each other through a long pipe of any diameter, the thing to do if we don't understand each other is to ask for the message to be repeated. Obviously, repeating the same message "wastes capacity"—that is, it reduces the amount of data that can flow through the pipe per unit of time. There are lots of ways in which the pipe might be noisy. For example, a small child may be beating on the pipe with a rock, or a baby may be crying nearby, or one end of the pipe may be placed in an acid rock dance club. The "signal-to-noise ratio" measures the extent to which messages will have to be repeated in order to be understood. This ratio can be improved by increasing the strength of the signal (shout louder), by reducing noise (give the baby a toy), or by increasing the sensitivity of the receiver (buy a hearing aid). Other things equal, the higher the signal-to-noise ratio, the more data can flow through a given transmission pipe. For a given signal-to-noise ratio, more data can flow if bandwidth is increased or if receiving equipment sensitivity is increased.

Digital Coding

One way to understand the tradeoff between bandwidth and other ingredients of communication links is to consider digital coding and compression. An analog signal is like watching someone waving a flag at the end of our 8-foot-diameter pipe. A digital signal is the transmission of a "one" when the flag is up, and "zero" when the flag is down. The real world that we see and hear is continuous, and analog signals tend to capture the data content of the continuity of reality. (In truth, analog representations, like digital ones, also fail to scan 100 percent of an image. Even a photograph consists of many small grains.)

A digital signal cannot fully convey reality, because it is always possible that, between the moment when the flag was up (signal = 1) and the next sample point, when it was down (signal = 0), the flag went through other motions. A mindless decoder would simply infer that the flag moved smoothly and directly from the up position to the down position.

A digital signal takes samples of the real world, or an analog image of it, at certain intervals, and reports the state of the world at the ends of those intervals. What happens between the sample points is then inferred or interpolated or "computed" by the receiver. (In some cases, the human eye or ear simply fails to notice that anything is missing.) In other words, the representation of reality between the sample points is invented. The ultimate objective frequently is to convince the viewer that the image received is "real." It may not be necessary, however, or even desirable to achieve the objective of re-creating subjective reality. For example, the Internet now transmits what are by conventional standards extremely poor quality video signals that appear in a small window on the computer screen. It may be that the content of such signals is (or will come to be) well suited to their form, so that higher technical quality (greater subjective reality) is not useful, or not worth its price. To take another example, many photographs look better and are worth more when some information is subtracted from them. Realism is but one aesthetic modality.

Whether the signal is analog or digital, devices are needed that can encode the signal and then, at the receiving end, decode the signal, "invent" intermediate realities, and transform it all into a form that we humans can recognize—an analog form such as a projection onto a video screen. Such devices or software—"converters" or "coder/decoders" (codecs)—are expensive, and the expense at the receiving end of a mass medium is of great significance because there are so many receivers.

In the 1930s, the cheapest way to get a video signal from a broadcaster to a viewer was to use analog scanning and coding devices to impress the sample information on a 6-megahertz carrier wave, and to employ corresponding decoding equipment (a television set) at the receiving end. Over the intervening half-century, it has become much cheaper to do the same thing in a different way, using digital rather than analog technology, and in some cases using much narrower bandwidths and different transmission media. The cost of computer power has fallen very rapidly for two generations. Indeed, the decline in price has been even more rapid than the falling cost of transmission pipelines. Consequently, it has become economical to save bandwidth by substituting computer equipment at the ends of the pipelines. Alternatively, the same bandwidth can be used to send a higher-quality picture.

The capacity of a transmission medium depends on the range of frequencies that it transmits, also called its bandwidth, along with other factors such as storage capacity and processing power. In a digital context, the term equivalent to bandwidth is "bit rate." For every bandwidth, measured in hertz, there is an equivalent maximum bit rate, measured in bits per second, other things equal. A conventional analog TV signal has a bandwidth of 6 megahertz; this means a certain broadcaster can emit signals in the range between, for example, 54.0 and 64.0 megahertz, which corresponds roughly with TV channel 2. The 6 megahertz of bandwidth used for content is equivalent to 180 megabits per second of uncompressed video data at standard sampling rates with current technology. Transmission standards plus bandwidth thus define channel capacity. (Standards prescribe the method of encoding and transmission.)

In broadcasting, only the bandwidth and specific frequencies permitted by the FCC can be used. TV broadcasters are each allocated 6 megahertz. Therefore, each can broadcast only one channel of (conventionally defined) analog television signals.

Digital Compression: Procrustes' Bed for Bits

If a video signal is digitized, it can be compressed, using computer algorithms, into less than 6 megahertz. That is, it can be transmitted on a narrower bandwidth with the same or better reception. For a number of years, operators of communication satellites and other high-capacity communication links have employed digital compression to increase the capacity of their facilities. Telephone companies that transmit large volumes of audio data through telephone trunks have done the same. It is not uncommon for satellites that distribute TV signals to cable headends and local broadcasters to have compression ratios of five or six to one. That means the information in five or six TV channels is squeezed into 6 megahertz. There is no one ratio of normal to compressed channels because the degree of compression that is possible depends on the amount of redundancy in the uncompressed signal. Channels with lots of action cannot be compressed as much as channels with less action. Material broadcast "live" in real time cannot be compressed as much as material that can be run several times through a processor.

Any form of coding that involves prearrangements between sender and receiver can result in compression. For example, suppose two people agree in advance on two signals; each signal "stands for" some longer message. The letter "A" will mean "the turkey is in the oven and will be ready by the time you are scheduled to be here." The letter "B" if transmitted will mean something else. In this example, a message that is 528 bits long ("the turkey is in the oven and will be ready by the time you are scheduled to be here" is 66 characters or bytes long, times 8 bits per character) has been compressed to just 8 bits ("A"). In general, the more the sender knows about the form and possible content of the message, the more succinctly the message can be coded, and the cheaper the necessary equipment. However, in the example only 8 bits are sent, rather than 528 bits. If each could be sent in one second, was the data rate 8 bits per second or 528 bits per second? This is the sort of ambiguity that confuses many discussions of bandwidth, communication capacity, and data transfer rates. (Indeed, from a strictly information theory point of view, the more the recipient knows about the range of possible messages, the less information is contained in any message. The message "A" conveys far more information in this sense if the set of possible messages includes A, B, C, D, and E than if it includes only A and B.)

Video Compression: Use Your Imagination

Television operates in much the same way as a movie film. There is a series of frames, each a fraction of a second apart, and each a frozen snapshot of the action seen by the camera. When the frames are played back at the same rate, the human eye is fooled into interpreting the result as a smooth, continuous moving image. Whereas each frame of a movie film really is a photographic snapshot, each frame of a TV signal is several hundred horizontal lines, each representing a scan across the frame. In both a video camera and a TV set, an electron beam or the equivalent literally moves extremely rapidly across the screen, line by line (or sometimes alternate lines), until it reaches the bottom, and then repeats the process from the top, frame by frame. A fax machine works in exactly the same way, but much more slowly.

Suppose there is a single frame (one "screenful") of video that must be transmitted through some communication channel, whether over the air or via cable. If the frame conforms to current broadcast standards, it is an analog image that has been scanned at up to 30 frames per second, with 480 lines per frame. Sending this signal in the usual way over the airwaves or on a cable occupies 6 megahertz of bandwidth.

The first step in the compression of such an image is to digitize it. That means taking samples, as described earlier, at specific points—snapshots of the state of the analog image at periodic intervals. One popular standard for digitizing calls for sampling 544 times per line. Each sample generates what is called a "pixel"—a small square or rectangle representing the information from the sample at a particular point (see Figure 9.3). The more sampling there is per line, the smaller the pixels, and the finer the "grain" of the resulting image when received. Nevertheless, the area of the original image that is sampled is smaller than the pixel to which it corresponds at the other end.

A standard analog 6-megahertz video signal does not have any fixed corresponding bit rate, because it is not digital. The precompression bit rate depends, in other words, on how many times per second

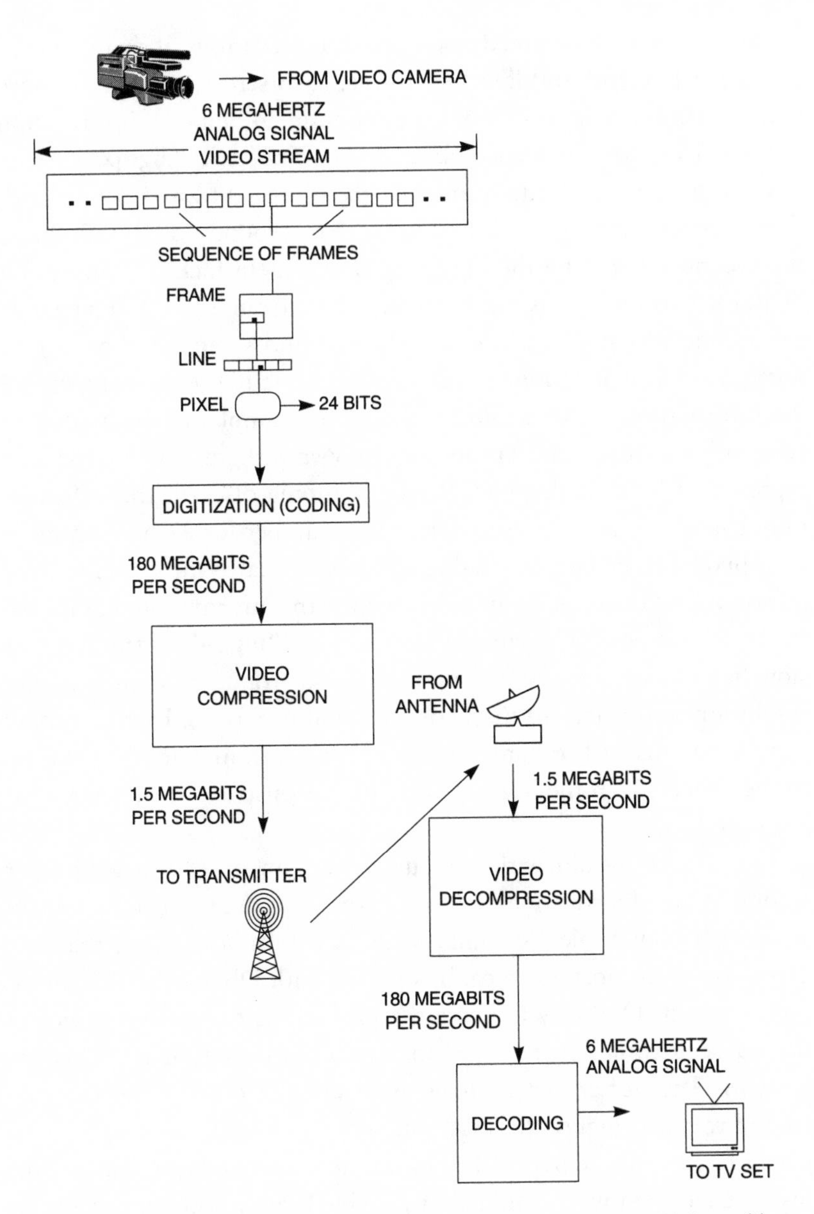

Figure 9.3 Digital video compression (C-Cube Microsystems, 1996; used by permission)

the analog signal is sampled or digitized. For example, if there are 544 pixels per line and 480 lines per frame, each screen consists of more than 250,000 pixels. Each pixel may contain 24 "bits" of information (luminosity, chrominescence, and so on). Thus each digitized frame has about 6,000,000 bits of data. If there are 30 frames per second, then there is a data flow or "bitstream" of around 180,000,000 bits per second emitted by the digitizing process. (In fact, uncompressed digital television signals can have anywhere from 30 to 270 megabits per second of data, depending on format and sampling choices.) As noted above, digitization of this sort has already greatly compressed the information in the analog image. For example, in the analog or conventional image, tint or intensity may vary slightly within the area represented by a single pixel. Digital scanning discards this information, usually by averaging to derive one number for each attribute of each pixel. (To be fair, the analog scan also discards information—the information "between the lines" as well as the information that is "between the frames." The current standard for "high-definition" television has twice as many scanning lines per frame, and thus higher resolution (information) than conventional television. Even a motion picture film discards the information between frames and leaves it up to the processor in the human brain to "reconstruct" or invent what has been lost.

What does one do with a digital bitstream of 180 megabits per second? One objective may be to transmit this data to someone's TV set over a communication link, such as a telephone wire, that can carry—say—at most 1.5 megabits per second. Obviously, 180 megabits won't fit. One way to solve the problem is to install a device (a hard disk or some other buffer) that stores information at the receiving end. The buffer can be filled with data at the 1.5-megabit-per-second rate (see Figure 9.3). At some point, the end user can begin to play it back at the "correct" 180 megabits per second rate. This works just like elevated water tanks that provide large volumes of water on demand at peak periods, and that are themselves fed continuously by pumps at relatively slow rates.

If electronic storage were very cheap and if there were no penalty for the delay involved, storage would be a good solution to the problem of transmitting 180 megabits down a pipe that is only 1.5 megabits wide. In many respects it is equivalent to sending a videotape or disk through the mail. It will not work, however, for real-time transmissions such as video conferences or live coverage of news or sports—unless the delay is very short.

Another way to solve the problem of fitting a 180-megabit data stream into a 1.5-megabit channel is to eliminate some of the bits. For example, one could send the frames in black and white rather than color, which would require fewer bits per pixel. A more imaginative way is to "code" the data. The message about "the turkey in the oven" was coded into the letter "A." Many people use numerical pagers in this fashion. Provided that the recipient of the message understands the code, very large volumes of data can be transmitted through a narrow pipeline.

Here is a way to understand the tradeoff between transmission capacity and coding capacity: suppose one were going away on a long business trip to Uzbekistan, where international telephone rates are extremely high and mail service is unreliable. It might be sensible to work out in advance with one's spouse and friends a rather elaborate code of the "A = 'turkey is in the oven'" variety, so that the length of calls back home could be minimized. How much time one should spend preparing such a code book obviously depends on how high the rates are. Time and effort spent on message compression would vary inversely with the cost of bandwidth.

Suppose part of the video frame is an image of a red apple. That means there are many red pixels in a row on each of several lines used to scan the image. Rather than sending information about each and every pixel, a computer can decide to send only the information necessary to *reconstruct* the image. Here the reconstruction requires not merely the interpolation of conditions between sample points, but the reconstruction of sample points themselves. Thus the stream of bits literally describing the image can be transcribed into one that permits the image to be computed at the receiving end. Instead of mindlessly sending 300 consecutive bytes, each of 8 bits, and each of which repeats "red dot," a message requiring 2,400 bits, the sending computer sends the information "red dot repeated 300 times" (26 bytes or 208 bits). The computer at the other end is programmed to interpret this and to reconstruct the image and to generate 300 red dots. In this way, the whole 180 megabits per second of "literal" information can be transmitted in a small fraction of a second on a 1.5-megabits-per-second channel—but only if the original series of images did not "really" contain 180 megabits per second of information to begin with, because of redundancy. A great deal of the advance in transmission technology in recent years has taken place because engineers have been able to use more powerful computer processors to engage in this coding and decoding of data, and thus to increase the apparent bit rate subjectively experienced by the user through a pipeline of given bandwidth. Figures 2.2 and 2.3 demonstrate the increasing power of microprocessors and memory devices.

To return for a moment to our culvert, imagine that there is a large block of ice at one end that must be transmitted to the other end. The ice can be chopped up and pushed through the pipe, or with enough deforming pressure it can even be squeezed through in one piece. But more ice can be shipped through faster if the ice is first melted. The volume of a given amount of water is greater in solid than in liquid form. Melting "condenses" the ice. Thus, more ice can be transmitted through the pipe with a given effort. However, one needs to install ice-melting equipment at one end and ice-making equipment at the other end in order to reconstruct the ice. The issue is whether it is cheaper to buy this equipment or to increase the pipe size.

Another common form of compression takes account of the fact that in a stream of video data consisting of a series of frames, or snapshots, many of the pixels in any given location on the screen are the same in adjacent or consecutive frames. Indeed, only in action sequences such as sporting events is there likely to be a big change from one frame to the next. In other sequences, the stream of video data can be greatly compressed by having the receiving unit "remember" the prior frame, so that the communication channel need contain only information regarding the pixels that have changed. This requires both computational power and storage facilities (a buffer or cache) at the receiving end. (Internet Web browsers, such as Netscape Navigator or Microsoft Explorer, use much the same technique. Copies of recently visited Web pages are stored in a cache on a local hard disk, where they can be called up, rather than requiring often slow retransmission of the data.)

The most popular current standards for video compression are ones established by a voluntary worldwide industry association, the Motion Picture Experts Group (MPEG), and adopted by the International Standards Organization (ISO). The first of these, MPEG-1, introduced in 1992, is tailored for use in connection with compact disks that contain video images. It calls for a data transfer rate of 1.5 megabits per second, the same as a T1 trunk telephone line, and provides video images of about the same quality as a VHS videotape. The more recent MPEG-2 standard has been adopted by the direct broadcast satellite industry and by the FCC for the next generation of digital television. MPEG-2 is designed to accommodate high-definition television with a wide screen such as those in a movie theater, accompanied by digital sound, as well as improved quality for "standard" definition television broadcasts. MPEG-2 will also be used for the next generation of compact disks (digital video disks, or DVD). The data transfer rate for MPEG-2 is variable, with a maximum of about 10 megabits per second. MPEG-2 uses "interframe" coding, transmitting only the pixels that have changed from one frame to the next. In addition, some redundant information within frames is removed, along with certain aspects of each image that have minimal visibility to the human eye, such as vertical and diagonal movements. In some theoretical circumstances, compression ratios of 200:1 can be achieved, though not with real-time processing. The practical range seems to have an upper limit of 30:1. Table 9.1 shows the transmission times for one minute of MPEG-2 video on various media.

MPEG-4, expected to become final in 1998, will provide for video transmission over narrowband channels—that is, telephone wires. The video image will be of corresponding lower quality, with fewer pixels per frame, and a data transfer rate of 48 to 64 kilobits per second. This compression protocol is intended for use with video bitstreams sent over Internet facilities and for video conferencing.

Current video bitstreams on the Internet, such as those supported by RealVideo and VivoActive, use advanced compression techniques such as "discrete cosine transform," "wavelets," and "fractals." These are all complex mathematical coding algorithms. The most interesting is "fractals," which involves the identification of small areas in a frame that are visually indistinguishable from larger areas in that frame or the next. The smaller area is transmitted and then "blown up" by the decoder to fill the larger area, reducing the required data

Table 9.1 Transmission pipelines with MPEG-2 compression

Connection	Top speed (kilobits per second)	Bandwidth (kilohertz)	Pages of text per minute	Time to transmit one minute of MPEG-2 video
Older PC modem	14.4	4	5 pages	100 minutes
Standard PC modem	28.8	4	10 pages	50 minutes
Digital telephone line (ISDN)	128	4	40 pages	6 minutes
High-speed digital telephone line (T1)	1,500	4	500 pages	1 minute
"Modem" for cable TV system	20,000	6,000 shared	3,300 pages	9 seconds
Fiber-optic backbone trunk (T3)	45,000	2,688	15,000 pages	2 seconds

flow. Even this level of sophistication is not sufficient to produce satisfactory video images over today's Internet. It is not certain that MPEG-4 will improve matters.

Compression provides no free lunch. Although it conserves bandwidth, it imposes costs at each end of the channel, which require computational equipment and software of sufficient speed and power to achieve the desired throughput. As already noted, when an image is sampled or digitized some information is lost; no matter how many sample points there are per inch, or how many pixels per frame, there is still lost information between the sample points. In many (but certainly not all) contexts, this represents a cost because the lower quality of the image has a lower value to users.

One of the most interesting dynamics of the possible "convergence" of video and telephone media is the tradeoff between bandwidth and coding expense. With unlimited free bandwidth there would be no need to compress signals, and both sending and receiving equipment would be much less expensive. When bandwidth has a price, there is always one or more particular combinations of bandwidth and terminal equipment that minimize the cost of the system. (There may be more than one if each produces a different image quality.) In a communication network that uses wires or fiber-optic cables, this tradeoff is internalized. The owner of the network chooses the combination of bandwidth and compression that minimizes costs, given current technology. But in the case of communications that use the airwaves, the bandwidth is determined by the FCC, and until recently did not have a price. (The government has begun in the last few years to use auctions to assign certain portions of the airwaves to certain users, but the licenses are not fully assignable, nor can they be used for anything other than their initially designated purposes.)

The Great Bandwidth Shortage

In recent years the government has greatly increased the quantity of over-the-air bandwidth available for video transmission. This means that the price of bandwidth will fall. Will the famous "bandwidth shortage" end? This is a tricky question. In one sense, there never was a technical shortage of bandwidth; getting more information could always have been achieved by using more sensitive and more expensive equipment. From this perspective, it makes no more sense to talk about a bandwidth shortage than to talk about an equipment shortage. On the other hand, the supply of spectrum, unlike the supply of equipment, has not been subject to market forces. The economic rents associated with broadcast licenses and the enormous arbitrage play represented by the cable industry strongly suggest that regulators have supplied far less bandwidth for video than a free market in spectrum would have done. As for the use of equipment to offset this shortage, it is telling that the FCC has always specified in exquisite detail how much information its broadcast licensees are permitted to transmit per unit of bandwidth. These standards are seldom changed. This is just the sort of action an effective cartel manager would take to prevent cheating on an output restriction.

The final element that can substitute for bandwidth is storage. Storage is available on a grander scale than the buffers holding a single frame or hard disks with an hour or two of video. It is important to recognize that much of what gets transmitted through communication links is material that can be stored and physically transported with little or no loss of value. This is especially true of information intended to entertain, but it is also true of "serious" information. While there are several Web sites where dictionaries reside, the same dictionaries are also available in CD form, for use with computers, and as old-fashioned books. A TV broadcast of a prerecorded sitcom could be sent to consumers through the mail as a videotape, just like a magazine subscription, presumably with little loss of consumer value. As with compression, there is a tradeoff between storage and transmission costs that helps to determine what gets stored and transported and what gets transmitted. If storage media were cheap enough, there would be no transmission, only transportation.

Of course, cost is not the only factor. Timeliness is often important. It certainly is with news and sports. Further, the form taken by transmitted information will generally be different from the form retrieved from storage, and this may affect consumers' willingness to pay for the information. The difference between a movie seen in a theater and the same movie on videotape illustrates the point. The Canadian communication theorist Harold Innes spawned a whole literature on the ways in which the medium conditions the meaning, significance, or value of messages. Or, as Marshall McLuhan put it, the medium *is* the message.

To further illustrate the tradeoffs between storage and transmission, if only whimsically, imagine that all of human knowledge as it exists each day could be contained, along with all the latest movie releases and daily soaps, in a single, inexpensive compact disk, delivered like a daily newspaper. Obviously, the demand for data transmission services would greatly decrease. Indeed, the only remaining transmission demand would be for real-time activities (news, sports, telephone conversations, and e-mail) and for the distribution of newly created knowledge so important that it could not await the next day's disk, such as storm warnings. Similarly, in bringing public libraries to many communities, Andrew Carnegie early in this century helped establish local storage devices for data, then a much cheaper way for readers to obtain information than the available transmission-based alternatives. There has always been a race between transmission and storage technologies, and it would be ironic if all the current focus on alternative transmission technologies and bandwidth issues were to be mooted by an unexpected breakthrough on the storage front.

Conclusion

A medium of communication reflects tradeoffs among bandwidth, processing power, and storage, among other factors. In addition, for any given combination of technical parameters, there are varied forms that the content of the medium may take.

Although bandwidth is often used figuratively to refer to the capacity of a communication medium to transmit information, it has a much narrower definition as the range of frequencies in use. The capacity of a medium is a function of bandwidth and other factors, not bandwidth alone.

Coding or compression is a way of saving bandwidth, or of crowding more information into a given bandwidth. The process of coding and compression takes place by digitizing analog signals and then discarding redundant information. Digital computers are very efficient, and growing more so, at coding and compression. The effect has been to multiply the potential capacity of existing communication facilities, such as telephone wires.

Storage can also conserve bandwidth, as happens when we watch a rented videotape instead of viewing the same movie using video-on-demand television.

The future of digital media, including television and the Internet, depends in part on what turns out to be the cheapest combination of bandwidth, storage, and processing that satisfies consumer demand for new digital services.

CHAPTER TEN

NETWORKS AND PIPELINES

The structure of a communication network affects its value no less than does the transmission technology itself. A conventional television broadcast is a one-way network that goes from one transmitter or a few transmitters to many receivers. A private telephone wire strung between two offices is a two-way network. Neither would be suitable to serve the purposes of the other.

Economies of Switching

Everyone in the world could be connected to everyone else by a direct private telephone connection (assuming there were enough copper in the planet). Such a network, however, would be monumentally wasteful, because most of the wires would be idle most of the time. To economize on the use of connecting wires, a telephone network takes advantage of the fact that most telephones are in use briefly, and not all at once. In every neighborhood, each telephone is connected, not directly to all the others, but to a local switch (see Figure 10.1). The switch can connect any telephone with any other local telephone. Further, the switch can connect any telephone with any long-distance company (each of which has in effect a number of local phones in the area served by each switch).

The use of switches drastically reduces the number of wires needed to ensure that anyone can call anyone else. Further economies are realized by measuring traffic flows and building switches that are only big enough to handle the maximum number of simultaneous calls that are expected to take place. If a local switch serves 10,000 telephones, typically only about 2,000 are likely to be in use at any given time, even at the busiest period. So the switch is only big enough to handle 2,000 simultaneous connections, plus a reserve. The numbers that one "dials" to initiate a call are simply instructions to the switch (itself a computer), telling it how to make the necessary connections.

A long-distance company faces exactly the same problem as a local telephone company in designing a network. It must find ways to connect thousands of local switches with one another. To do this, long-distance companies connect local switches to their own regional switches. Calls within the region pass from a local switch to the regional switch to some other local switch. Calls between regions pass either through trunk lines directly connecting regional switches, or to yet another switch that accomplishes the same thing. Some long-distance networks have four layers of switching.

In designing interactive networks there is always a tradeoff between switching costs and the costs of transmission links (trunks). If MCI had, say, twenty-five regional switches, it would not be preposterous to imagine connecting them all with each other directly, rather than connecting each of them with a central switch or switching hierarchy. Whether it makes sense to do so simply depends on the relative costs of the two forms of connection. Running wires or cables across the country to connect each of twenty-five switches with each other seems likely to be very expensive, relative to connecting each to a central switch. But if the connections are made through a communication satellite, rather than wires, such an arrangement does not seem so far-fetched.

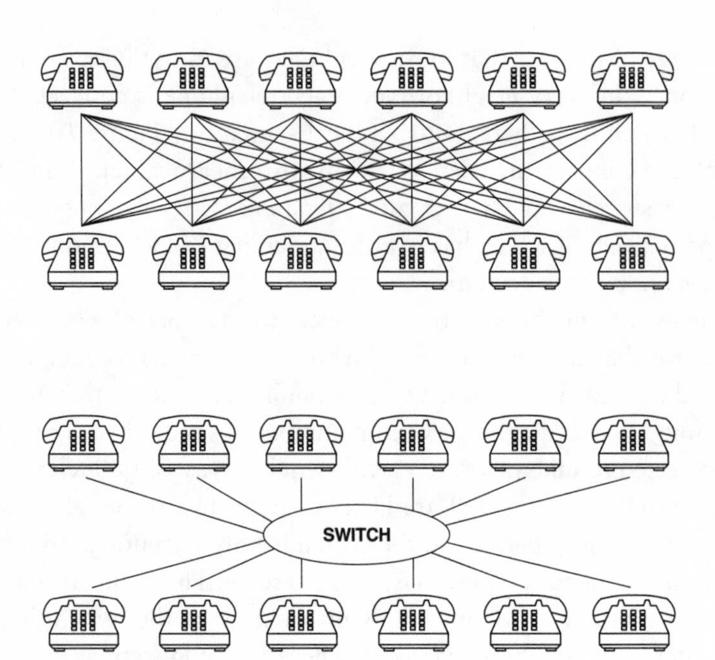

Figure 10.1 Economies of switching

In addition to switches, local and regional, the telephone network has three other important components: the equipment such as telephones or modems that is owned by the customer and that is connected to the telephone lines; the "local loop"—typically a simple pair of copper wires—that connects each telephone to a local switch; and the trunk transmission paths between the switches. Together these four components define and limit the capabilities of the network (see Figure 10.2).

Rapid advances have been made in trunk and switching technology. Most telephone trunks are now composed of fiber-optic cables, which have enormous capacity (measured in trillions of bits per second) and hence very low unit costs at large scales. Similarly, most switches are now high-speed digital computers capable of doing far more sophisticated tasks than simply switching calls. Because these switches use the same components and technologies as other computers, switching costs have been falling. Many telephone customers own digital communication devices, such as high-speed computer modems, the cost of which has declined as well.

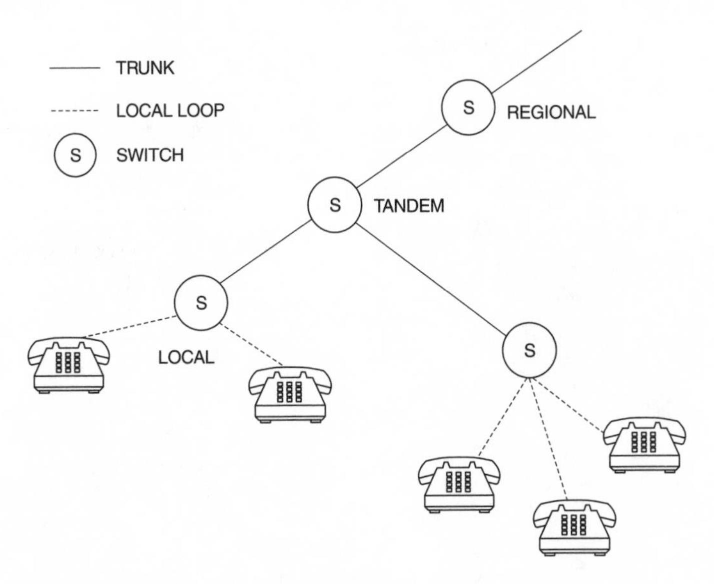

Figure 10.2 Telephone switching hierarchy

Local Data Transmission

The bottleneck in the telephone system is the pair of copper wires linking each telephone to the local switch. This is essentially the same transmission technology that was in use a century ago. From the local switch on up, the telephone network is a modern digital-switched system. But the local loop remains a slow, narrowband analog link, a nineteenth-century telegraph wire. (In the nineteenth century, at first both telegraph and telephone connections were made with a single wire, using the ground as the return connection. Later, for improved signal quality, both adopted wire pairs, or loops.) And yet every communication that uses the telephone network, including the vast majority of Internet transmissions, must pass through this wire. One reason for the persistence of this old-fashioned technology is that the local loops make up the lion's share of the local telephone company's capital investment (see Figure 10.3).

A television broadcast is point-to-points (one-way, unswitched, broadband, analog); a telephone network is point-to-point and back (two-way, switched, narrowband, digital in the middle, analog at the ends). There are other kinds of networks beyond these two important

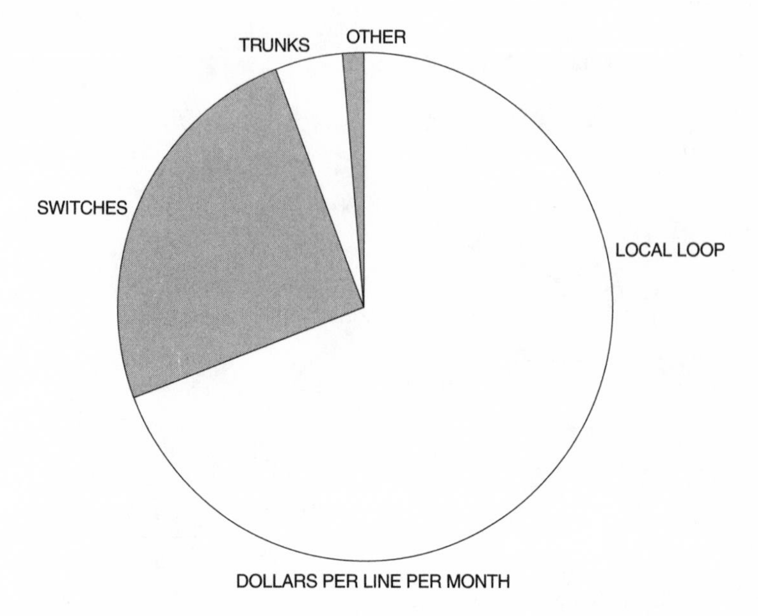

Figure 10.3 Representative costs of local telephone components (Hatfield Associates, Inc., 1997)

ones. For example, a cable television system is very much like an over-the-air broadcast: it is one-way, unswitched, broadband. Like the phone system, it typically has digital trunks but analog connections to the home. The only real differences between a cable system and a broadcast (aside from the fact that one is wireless) are that a cable system is multichannel and can more easily deny service to people who don't pay, simply by unhooking them from the cable. But broadcasters could in principle do much the same thing, using scrambled signals.

In contrast, the two-way local analog telephone system consists of a switch (typically, a computer), trunk connections between such switches, and a separate pair of copper wires leading from the switch to every telephone in the area. The amount of information that can be carried on the copper wires is limited by technology. An ordinary voice channel has a bandwidth of 4 kilohertz, and the telephone system was originally designed to provide that service. Consumer modems currently provide data at speeds up to 56.8 kilobits per second. They do this by means of compression—not every bit entering the modem gets sent out, but the original input can be reconstructed at the other end. As compression technology improves, analog modem speeds may continue to grow, although 35 kilobits per second on voice-grade telephone lines was once considered the theoretical maximum.

According to Shannon's Law, the maximum number of bits per second that can be sent accurately over a noisy channel is limited by the bandwidth of the channel and the signal-to-noise ratio. Noisy channels transfer less information than do quiet ones. This is why a cable television system can send more data in a given bandwidth than a wireless system. Over-the-air broadcasts have a bigger problem with noise. Noisy channels have less data transfer capability because the signal must include enough redundancy (extra data) to permit the receiver to accurately interpret the message. The simplest sort of redundancy is the repetition of a message.

In 1997 U.S. Robotics and Rockwell each introduced (incompatible) so-called 56-kilobit modems. To do this they used certain technical characteristics of the voice-grade channel to reclassify what had been seen as random "noise" into a deterministic or predictable effect; the result was to increase the signal-to-noise ratio and hence the maximum data throughput.

U.S. Robotics and Rockwell each rushed to bring out their innovative modems, hoping that the market would make one or the other the clearly preferred standard. As it turned out, it was a dead heat, and a middle path has now been established in the international standards organizations, so that the two flavors will no longer be incompatible. The point is that the technology got to market and into the hands of consumers in a few months. In spite of confusion and initial incompatibility, consumers benefited almost immediately from the new technology. It took only two years or so for the dust to settle. With the government in charge, in contrast, it took ten years to settle on a digital television standard, and because of the nature of the standard it will be more years before any significant number of consumers have access to the technology.

A different strategy for local data transmission is to change the telephone network from an analog to a fully digital system. One of the shortcomings of the analog system is that no matter how many switches are involved, a telephone connection is just that, an open connection dedicated to the calling parties and not available to anyone else, whether or not the parties happen to be talking. This "wastes capacity" or creates an opportunity to conserve capacity by eliminating redundancy.

In a digital system an "open line" between an America Online user and the local "switch" does not tie up capacity on the switch or on the connecting trunks. Instead, the switch pays attention only to the bits actually sent or received by the user. The copper wire twisted pairs used for telephone lines to the home can also carry digital signals, under some proposed technologies up to about 6 megabits per second. This is adequate for a single compressed TV channel. Digital engineering can in effect greatly increase the capacity of the existing telephone plant. (Nevertheless, it probably would not make sense to do this for voice traffic alone, even though the telephone network is digital starting at the local switch, because of the cost of the equipment that would be needed at every telephone to digitize the voice signals.)

U.S. telephone companies have for several years offered a digital service for computers called the Integrated Services Digital Network (ISDN). ISDN offers data speeds of 56 kilobits per second on each of two data channels. At the moment, ISDN is the fastest generally available consumer data service—at speeds equal to the Internet backbone speed in 1975. Ironically, even though ISDN service does not impose on the local switch when it is not actually sending or receiving bits, telephone companies typically charge by the minute for its use, even when they do not charge by the minute for ordinary local telephone calls. Go figure.

Packet Networks

The interconnected networks that today make up the "backbones" of the Internet are very different in structure from the telephone model. They are entirely digital. Communications from one point to another do not utilize "connections" or open channels between the points. Instead, each message is processed by the computer at its point of origin into bundles of bits called "packets." Each packet (of perhaps a few hundred bits) contains a small part of the message content plus a good deal of "overhead" information, including the address to which the message is going, the sender's address, and error-correction data.

Each packet is transmitted to the nearest "router" (see Figure 10.4).

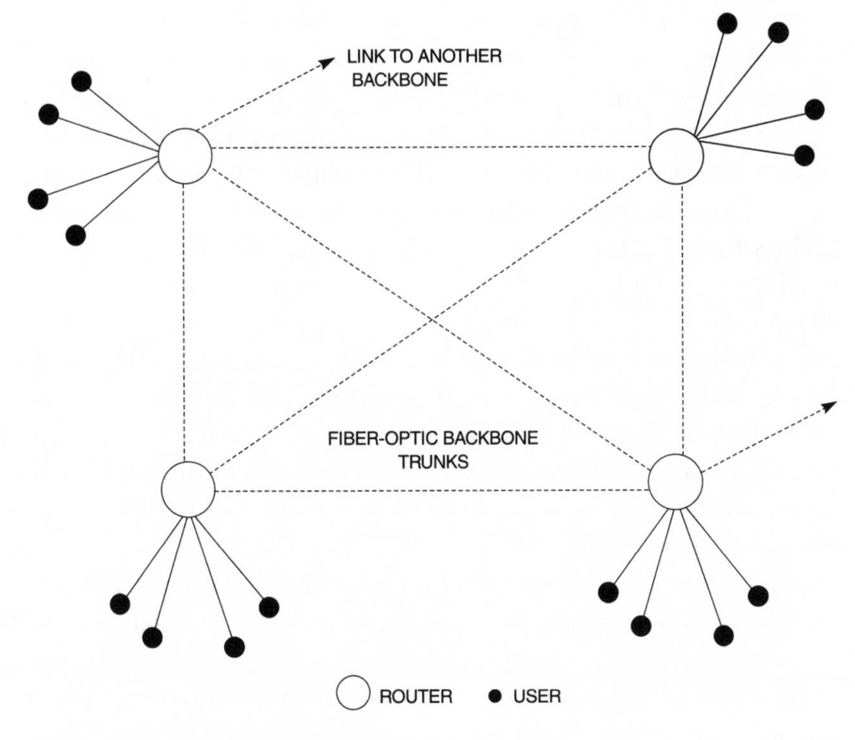

Figure 10.4 Internet configuration

A router has much the same function as a switch, except that it does not establish connections. Each router in the system is connected to several other routers. As packets arrive at a router, the router decides by which transmission path to send them onward toward their final destination. Typically, many paths are available. The packets that arrive at the destination usually arrive out of order with delays of variable length. As each packet arrives, notification of that fact is sent to the originating computer; otherwise, the originating computer keeps sending the same packet again and again. The destination computer reassembles the message from these packets. Unlike a telephone conversation, in a packet network there is no real-time linkage between sender and receiver. (If the network is fast enough, of course, it may appear to the users to be a real-time connection, just as the rapidly advancing frames of a movie or TV screen deceive the viewer's slow-reacting eye.)

One advantage of a packet network is that it makes efficient use of trunk lines. In a conventional telephone network a great deal of the capacity of trunks is reserved for connections that do not need such capacity, or don't need it at every instant. A packet network has no connections—no permanent exclusive path linking sender and receiver. Trunks in a packet network can be filled to the brim with packets traveling between any two points on the entire network. A disadvantage of a packet network is that, depending on the capacity utilization rate of its routers and trunks, portions of messages may arrive with considerable, and random, delays. The delays are reduced if elements of the network are not working at capacity. Capacity is determined by the bandwidth of the channels (and the routers) given the relevant transmission standards (the technical descriptions of messages and protocols). So once again, the quality of "reception" is affected by bandwidth.

Packet networks were originally designed to suit military needs for reliability and survivability. As long as each of the routers is connected to several others, a packet can find its way from one point to another even if several trunks are disabled on the most direct path. An ordinary switched telephone network also has alternate paths by which connections can be made in the event that a particular trunk fails.

Waiting for Godot

Delay (or what the engineers call "latency") is an important characteristic of computer networks and the equipment that we use to connect with them. An ordinary over-the-air television broadcast of a live event does not reach our eyes instantaneously. There is a slight delay. The electrical signals must travel from the camera to the studio and then to the transmitter and then to our TV set. These signals can go no faster than the speed of light. If the signal is processed at any point on its journey (for example, compressed and then decompressed), a further delay is introduced because the processing must take place in real time.

Delay is no problem for broadcast television because most viewers are unaware of it. (They can be made aware of it only if, for example, they are eyewitnesses to the events being broadcast, as in a sports arena, where fans sometimes watch small portable televisions.) In other contexts, however, latency is a significant problem. For example, when telephone calls are transmitted over certain satellites, there is a perceptible (500-millisecond) delay attributable to the speed of light. This delay can make ordinary conversations extremely awkward. In contrast, the delay attributable to the finite speed of light is only about 100 milliseconds for a low Earth orbit satellite or a 3,000-mile fiber-optic cable.

In any communication system that involves human interactions or human-machine interactions, perceptible delay reduces the value of the communication. Imagine channel surfing on a future Internet-based TV system that has a five-second latency factor. Each time one changed channels, one would have to wait an intolerable five seconds for the new channel to appear. This latency factor is quite independent of the bandwidth of the channel through which the video data are flowing. Apparently there are no studies of what sort of delay is acceptable, or perceptible, to humans in such circumstances, but a working standard of 100 milliseconds (one-tenth of a second) is generally used. Many modern communication networks, including the Internet, cannot meet this standard, especially with consumer-type processing and modulation equipment.

Packet switching is one reason for latency on the Internet. Packet networks are "connectionless"—there is no dedicated channel between sender and receiver. Instead, the transmission is divided into small pieces—packets—that must each find their way through the network and then be reassembled. Packets can be delayed or lost at each node in the network; congestion increases the number of nodes the average packet must transit; and reassembly of the packets must take place in real time. In two-way or interactive communication, latency can reduce quality. The long delays often experienced by Internet Web browsers are often blamed on latency in backbone networks, but they are often from other causes, such as slow or crowded servers and noisy local loops. In any event, pure network latency apparently did not increase on the Internet until 1997 (see Figure 10.5, based on the "Internet Weather Report," showing the average time required for a single packet to travel from Austin, Texas, to San Francisco, Califor-

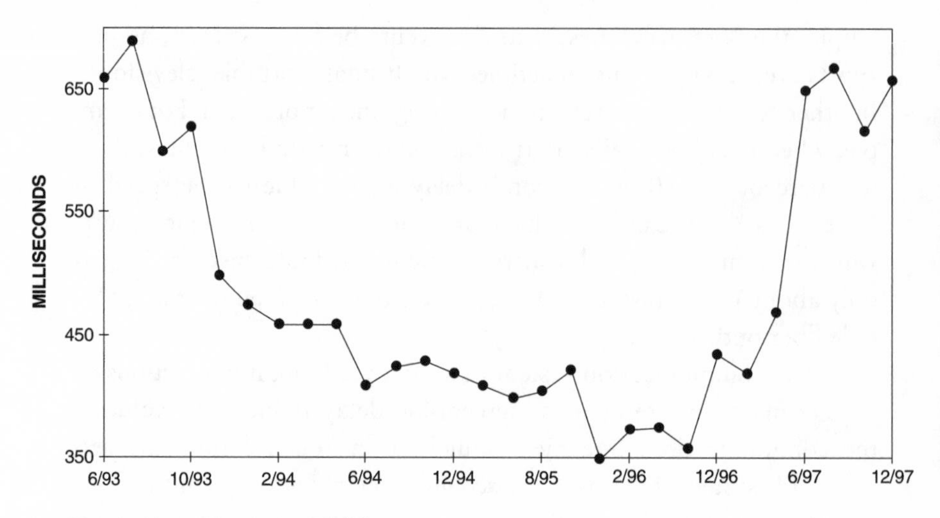

Figure 10.5 Worsening Web latency

nia, and back over the Internet). A Web measurement service called Keynote has measured latency in downloading a given number of bytes from forty commercial Web sites. For example, these forty business and commercial sites delivered home pages in 32 seconds on average during the week ending October 31, 1997. The best was Charles Schwab at 9.70 seconds, and the worst was 73 seconds. Keystone measurements cover the period beginning in late 1996 and show substantial average latency that varies greatly over time and by backbone, but no worsening trend.

Latency on the Internet is a special problem for "streaming" data used to transmit TV pictures. Among the solutions to this problem is a procedure called Resource Reservation Protocol (RSVP). This protocol attempts to reserve capacity in the Internet in a way that simulates the connections in switched telephone networks, as illustrated in Figure 10.6. In the upper panel of Figure 10.6, current IP packets are seen to take various routes through the Internet, arriving at their destination out of order. The process of reassembly in the correct order, and the wait for the slowest packet, introduces latency. In the bottom panel, a special RSVP router in effect reserves bandwidth for data streams and assures sequential arrival of the packets. Whether this will be a useful long-term solution remains in doubt.

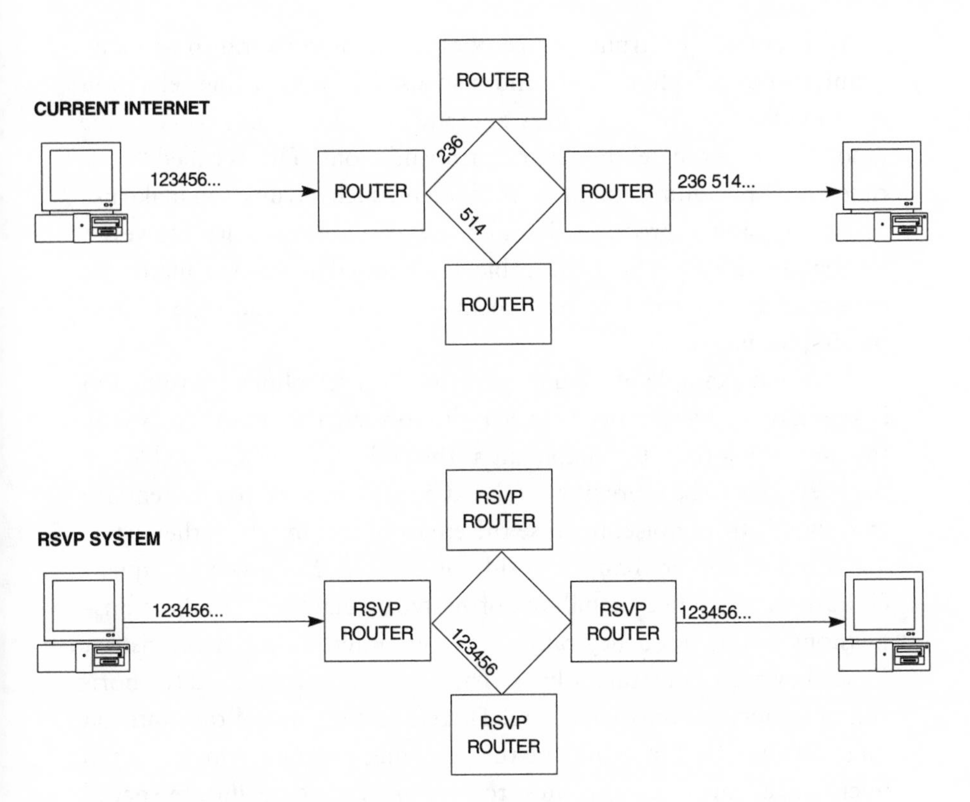

Figure 10.6 RSVP "connections"

Heavy Lifting: Trunk Technologies

Here we describe the technical features of several approaches to data transmission—coaxial cable, microwave radio, optical fiber, and communication satellites. In later chapters we will explore these technologies from a different angle, as business strategies.

Tremendous progress has been made in the development of communication trunks, defined either as physical connections between switches, or more generally as transmission paths that serve multiple users simultaneously. A "simple" length of twisted copper wire, used to connect a single party to the local switch, can with appropriate equipment and conditioning, and if it is not too long, be transformed into a link with trunk-like capacity, in the range of megabits per second.

An example of a trunk is a coaxial cable, used to transport many simultaneous telephone calls and also used in most cable television systems. Such a trunk requires equipment at both ends to combine (and then to untangle) the separate transmissions. This is called "multiplexing." By adding still more equipment, such a trunk can make use of compression approaches that increase available capacity. However, the bandwidth at which the cable is operated is itself a matter of choosing the appropriate equipment—the number and type of amplifiers, for instance.

A second example of a trunk often used for telephone transmission is a pair, or a series, of microwave radio towers. The radios do exactly the same thing that the equipment at the ends of the coaxial cable do, but they have to be more powerful and better equipped to differentiate true signals from noise, because the environment in which they operate is much noisier than the inside of a cable. Microwave frequencies are in the range of billions of hertz, or gigahertz. Radio transmissions in this frequency range have two interesting characteristics. First, they travel in straight lines. They do not curve around the horizon or bounce off the ionosphere. Therefore, the transmitting antenna must be literally "in sight" of the receiving antenna. On terrestrial microwave transmission routes, towers of average height are spaced about twenty miles apart. All the relay towers between the end points simply receive, amplify, and resend the transmissions from neighboring towers.

The second feature of microwave transmissions is that they are readily absorbed by water, where they are dissipated in the form of heat (which is how a microwave oven works). A heavy rainstorm can easily disrupt transmissions between two microwave towers. Many a telephone conversation has thus fallen as warm rain on the western plains. The higher in the gigahertz band the transmission, the more of a problem there is with water. At frequencies as high as 30 gigahertz, the antennas must be no more than a few miles apart, and even the leaves on trees can seriously disrupt communications.

The newest form of trunking takes advantage of advances in optical fiber. A fiber-optic trunk consists of one or more extremely thin, extremely long strands of optical-quality glass. Such fibers can now be made with very few impurities that could absorb or scatter the light. The frequencies used in a fiber-optic cable are in the range of visible light. This cable has unlimited capacity relative to present or foreseeable demand. While the theoretical limit on the capacity of the fiber may be enormous, the actual capacity is determined by the equipment at the two ends, and the amplifiers along the path. Fiber-optic bandwidths are measured in terahertz (trillions of hertz). By comparison, the current practical limit on bandwidth for a coaxial cable is around one gigahertz. Fiber-optic cables are now the chief means by which long-distance phone calls are trunked between switches. The country is criss-crossed by many thousands of miles of buried fiber-optic cable belonging to long-distance and local telephone companies. (The several dozen Internet backbone providers also use fiber networks.) The theoretical capacity of these trunks is many times greater than the present demand for long-distance service, and their actual capacity is constrained by the limitations of the electronic terminal equipment. Thus, while long-distance telephone companies and Internet backbone providers own fiber networks with enormous theoretical capacity, they have found themselves struggling to upgrade the actual capacity of their systems to meet growing demand.

Communication satellites are microwave towers high in the sky—22,300 miles high. Because they are so high, they are in "line of sight" to about one-third of the globe. And because the transmission path between a satellite and the Earth is more vertical than horizontal, it passes through less attenuating water vapor and rain than would a terrestrial microwave system. Traditional communication satellites are 22,300 miles up because that is the only altitude that permits them to orbit the Earth at the same rate the Earth turns. As a result, "geosynchronous" communication satellites appear to hover permanently in one spot in the same plane as the equator (see Chapter 8). This is an advantage because a large part of the cost of a satellite system goes into the many receiving antennas on the ground. Dishes for geosynchronous satellites can be simple and inexpensive if they need only point at the same spot in the sky at all times. But if the dishes have to move, tracking a satellite across the sky, they are much more expensive to build and to maintain.

One can think of the transmission path from an Earth station to a satellite and down to another Earth station as a trunk, just like a series of microwave towers. While a signal traveling at nearly the speed of light takes no perceptible time to travel 3,000 miles from New York to

San Francisco on a fiber-optic cable or a microwave system, the same signal takes a very perceptible half-second or so to traverse a satellite that is 22,300 miles away. Since nothing can travel faster than the speed of light, there is no remedy for this delay, which many people find annoying in telephone conversations and which also disrupts other real-time transmissions.

Communication satellites have other interesting characteristics. Among them is the potential to offer direct communication paths (whether messages or actual connections) from each telephone to every other telephone. A satellite simply picks up messages that are beamed up to it, processes and amplifies them, and rebroadcasts them over a very wide area. In principle, such a satellite could pick up messages from satellite dishes at every home and office, and rebroadcast the messages throughout its coverage area. Of course, only the intended recipient would be able to decode the messages addressed to him or her. Whether it would ever make sense to actually use this method to communicate is a matter of economics.

The Last Mile: Local Loop Technologies

Trunk communication technologies can be applied, on a smaller scale, to local loops, sometimes called the "last mile." For example, the hypothetical geosynchronous satellite system just described could supply all local loops and requires no trunks and no switches. Local loops can also be constructed from coaxial cable. The same cable now used to deliver cable television signals can be turned into a two-way broadband local area network (LAN). Microwave radio offers another method to construct local loops. With an appropriate antenna (not unlike the new 18-inch DSS satellite antennas), homes and offices could pick up local transmissions from cell-based microwave radio systems now in operation or under development (see Chapter 13). Any of these methods could replace copper telephone wires with a much greater bandwidth.

Of course, the capacity of the copper wire local loops can be increased by piling on appropriate equipment at the end points. For example, a so-called high-speed T1 connection, which is used for communication between large computers or to link one computer network to another at the rate of 1.5 megabits per second, is nothing more than a pair of copper wires to which special equipment is added at the sending and receiving ends and at the telephone central office. The reason we don't all have such connections is not that it would be necessary to tear up the streets, but that the terminal equipment in question is too expensive to be justified by present consumer demand. Telephone engineers have developed a variety of digital subscriber lines (DSL) that squeeze broadband signals onto existing copper wires.

Many local telephone companies are experimenting with these technologies, chiefly "asymmetric" DSL (ADSL). Asymmetric indicates more data flows in one direction (downstream, to the user) than the other. There are two basic difficulties that such technologies must overcome. First, they operate reliably only over short distances. Many residential telephones are several miles from the nearest switching office, which is too far for current DSL technology. Therefore, to implement DSL, telephone companies probably will have to do the same thing that cable television companies must do, which is to construct fiber-optic links to neighborhood nodes. Second, DSL is expensive—about $1,200 per subscriber, the bulk of which is for equipment at the ends of the loops.

Figure 10.7 illustrates three local loop technologies. The top panel shows today's standard connection (POTS, or "plain old telephone service") from households to the Internet. The household has a modem with a capacity of up to 53 kilobits per second (nominally 53 kbps) in each direction, connected by copper telephone wires to the local telephone central office or switch, which in turn is an outpost of the international core telephone network. Somewhere else, an Internet service provider has a local "point of presence," a room full of modems of similar capacity, each connected to its local central office by copper wires.

The middle panel of Figure 10.7 shows the same function accomplished by an ISDN digital connection. The same copper wires are used, but the modems are replaced by digital subscriber line codecs, which must be at each end of each local loop. These devices permit the establishment of two 64-kilobit-per-second digital channels, which may be combined, plus a 16-kilobit auxiliary channel, often used for voice. The transmission is digital only on the local loop; the core network remains analog.

The bottom panel of Figure 10.7 shows an xDSL connection. (The

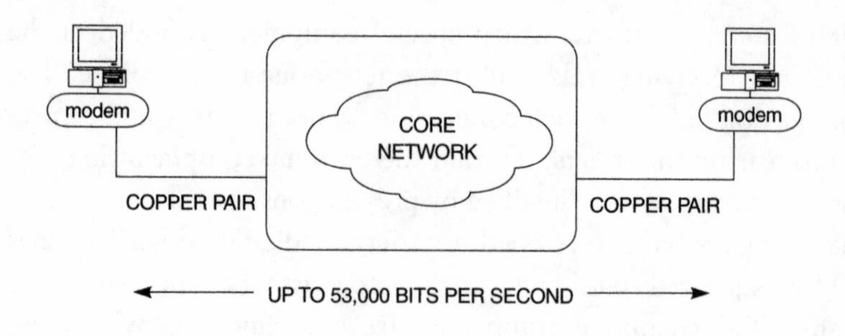

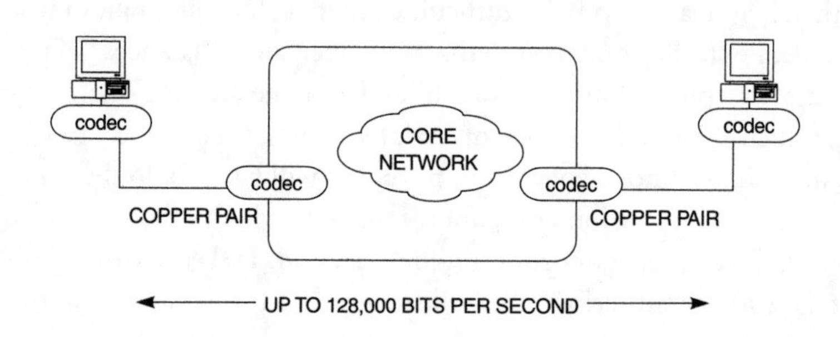

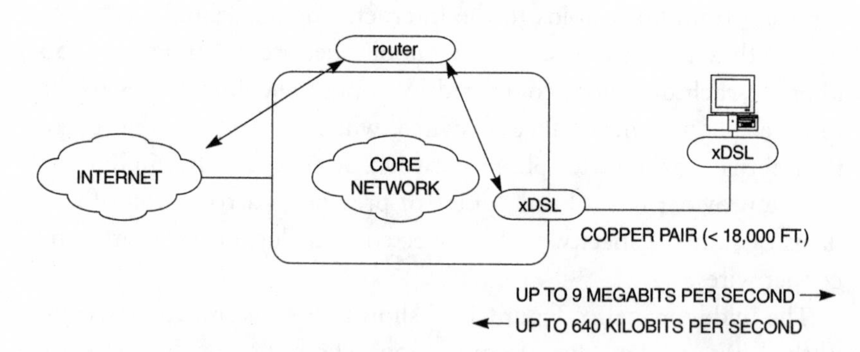

Figure 10.7 POTS, ISDN, and xDSL connections

Table 10.1 Last-mile technologies for Internet connections

Name	Meaning	Data rate	Mode	Applications
V.22, V.32, V.34, etc.	Voice band modems	1.2 to 53 kbps	Duplex	Data communication
DSL	Digital subscriber line	160 kbps	Duplex	ISDN service, voice and data
HDSL	High data rate DSL	1.5 mbps (2 loops)	Duplex	T1 trunk line
		2.0 mbps (3 loops)	Duplex	
SDSL	Single line DSL	1.5 mbps 2.0 mbps	Duplex Duplex	Same
ADSL	Asymmetric DSL	1.5 to 9 mbps 16 to 640 kpbs	Down Up	Internet access, VOD, one-way video, multimedia
VDSL	Very high data rate DSL	13 to 53 mbps	Down	Same plus HDTV
		1.5 to 2.3 mbps	Up	

illustrative numbers refer to ADSL.) The same copper wires can be made to carry up to 9 megabits per second downstream to the user if the user is located no more than a few hundred feet from the local switch, and 1.5 megabits per second at distances of around 18,000 feet. The capacity of the line in the opposite direction ranges from 16 to 640 kilobits per second, again depending on distance. In contrast to the preceding connections, however, xDSL transmissions are (or can be) routed to bypass the conventional switch to a server or router, and thence to the Internet.

Numerous digital subscriber line technologies could use existing copper telephone wires to the home (see Table 10.1). Of these technologies, telephone companies are currently focusing on ADSL.

Digital subscriber lines, like cable modems, have a range of data rates in theory and in practice. A cable modem of the sort introduced initially can receive data at up to 10 to 30 megabits per second. But the capacity used for this purpose is shared, and therefore the actual data rate in practice depends on the number of users sharing the capacity at each node, the amount of noise or interference on the cable, and many other factors, just as the water supply loses pressure when everyone turns on the tap. Similarly, in the case of DSL, even though each local loop channel is nominally dedicated to a user, many factors akin to sharing affect actual capacity. One reason for the focus on ADSL by telephone companies is that the electrical characteristics of bundles of copper pairs limit the return capacity of a digital line in the bundle. Also, DSL transmissions are very distance-sensitive. Although the theoretical limit is 18,000 feet from the central office, attenuation problems can crop up well before that limit is reached, depending on local conditions, which is why ADSL transmission capacity is usually given as a range from 1.5 to around 6 megabits per second. In sum, not only are the costs of and demands for consumer digital services unknown, it is not even easy to gauge in advance the capacity of the facilities that would be installed. Finally, if data were received at anything like the theoretical maximum rates, either for ADSL or for cable modems, it would overwhelm the serial ports and internal connections (busses) of most personal computers.

As noted, cable television systems can provide local loops, both for two-way switched telephone and narrowband data transmissions, and for broadband signals addressed to particular households. Cable companies can supply "virtual" high-speed local loops, but whether they can do so at a competitive cost and market the resulting service is a real question. So far the cable industry does not seem prepared to gamble heavily on any of these possibilities.

Processing

The electronic (or, in a few years, the optical) equipment at the ends of transmission lines constitutes the "processing" element of the bandwidth-processing-storage equipment tradeoff. There is a bewildering variety of microelectronic technology involved in these devices, although digital signal processors are currently the most popular. Digital signal processors are chips that specialize in real-time manipulation of data flows. They evolved more or less in parallel with microprocessor chips such as the PPC, Pentium, and RISC chips. Moore's Law seems still to apply: the amount of processing power (measured in "flops"—floating point operations per second) that can be purchased with a dollar doubles about every eighteen months or two years. Current processors perform at megaflop levels, and those on the drawing board will perform at gigaflop levels.

Convergence?

The television business and the telephone business are sharply defined today by regulatory boundaries and choice of technology. Before passage of the 1996 Telecommunications Act telephone companies didn't offer TV signals because they had not been awarded broadcast licenses or permitted to own cable television systems. (Actually, AT&T was a pioneer radio broadcaster, but as noted in Chapter 4 it divested its radio interests in 1926.) TV broadcasters don't use the airwaves to offer telephone service because, at least until very recently, regulators, both state and federal, would have prevented them from doing so. Cable television companies haven't offered telephone service because, until the Telecommunications Act of 1996, state laws made it impossible. These sharp legal boundaries may be about to disappear, to be replaced by economic boundaries perhaps no less sharp.

Technologically, telephone companies have offered narrowband two-way switched communication services, while video broadcasters of all sorts have offered broadband one-way service. Telephone companies and broadcasters both use the airwaves as well as "wires" (copper, coaxial cable, and optical fiber). In the world that lies just a few years ahead, these technology-based boundaries could melt away if a single, two-way digital network turns out to be the most efficient way to deliver television and other traditionally one-way services. But although this outcome is possible, it seems most unlikely. To be sure, other boundaries based on regulations, corporate culture, and consumer demand will remain. For example, it may turn out that a one-to-one relationship between commercial organizations and particular communication technologies happens to be the most efficient way to run such businesses.

The major reasons that are usually cited for the possible disappearance of distinctions between broadcasting and telephony include (1) the use of digital transmission methods for television as well as for data communications, (2) use of the same terminal, either a TV set or a personal computer, to receive, process, and display both video and data images, and (3) the so-called hybrid fiber-coaxial (HFC) network, a new technological standard for local telephone and cable services. (See Chapter 7.) Figure 10.8 illustrates this new technology.

Of course, neither the transmission path nor the terminal equipment is yet truly equipped to supply both broadband video and switched data service well. Present-day television screens are only marginally capable of displaying a readable Web page, for example, and the packet-switching system used for the Internet is not ideal for "streaming" data, such as real-time video. Although telephone companies and cable systems may both use fiber-node architecture, that alone is not sufficient for telephone companies to supply video signals to the home, nor is it sufficient for cable companies to offer services such as two-way Internet access or voice telephone service. Convergence is a possibility, not a reality, because it is still much too expensive. One way to think about the future is to ask which technology is most likely, at scale, to reduce the costs of these services to levels consumers will find attractive. Table 10.2 summarizes the various access routes for Internet connections and their costs. There is no evidence that consumers will be willing to pay even modestly for any of them.

Just as the choice of technology for a transmission path (paired wires, coaxial cable, fiber-optic channels, satellite channels, over-the-air broadcasts) is a matter of economics (and regulatory boundaries), so the success or failure of communication technologies—and companies—depends on their comparative economics. For example, under the current plan, terrestrial broadcast stations making the transition to the new digital spectrum assignments would be able to use the airwaves for any purpose, not just advanced television broadcasting. But under the 1996 Telecommunications Act, broadcasters are supposed to pay for the spectrum if it is used for purposes other than free television broadcasts. It is possible that the capacity made available to broadcasters in this way will enable them, either individually or in combination, to offer communication services that compete with what today is called cable television or cell phones. It is all a matter of capacity and cost per bit per minute for a given service.

Walter S. Ciciora (1995, 59) makes an interesting comparison of digital video and digital voice services.

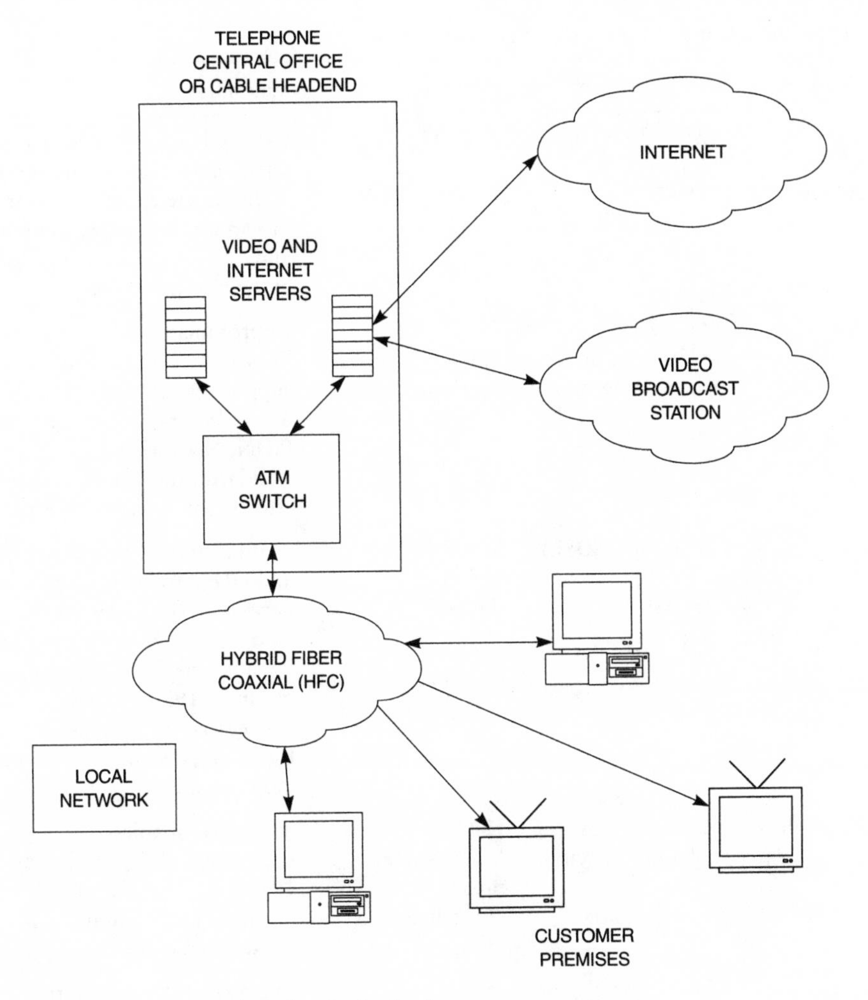

Figure 10.8 Last-mile technologies for Internet connections

People watch video for hours and video requires mega (millions of) bits per second. People talk on the telephone for minutes and voice requires only kilo (thousands of) bits per second. The ratio of the number of bits needed for telephony compared to the number of bits needed for video is in the range of hundreds of thousands. If "bits are bits" and their transport and switching comes at a given cost, either digital video will be incredibly expensive and digital voice affordable or digital video will be affordable and digital voice essentially free!

Table 10.2 Major end-user Internet access technologies

Technology	Downstream	Upstream	Summary
POTS (analog voice telephony)	28.8–33.6 kbps (56 kbps in 1997)	28.8–33.6 kbps	94% of homes have regular telephone service; requires no additional telephone company investment and only a computer and (inexpensive) analog modem at the user premises.
ISDN	56–128 kbps (230 kbps under development)	56–128 kbps (230 kbps under development)	Approximately 70% of access lines are now capable of supporting ISDN, but less than 5% of Internet subscribers use ISDN. New pricing, standardization, and marketing efforts may increase penetration.
xDSL	384 kbps (SDSL) 768 kbps (HDSL) 1.5–8 Mbps (ADSL)	384 kbps (SDSL) 768 kbps (HDSL) 12–500 kbps (ADSL)	Significant deployment of SDSL and HDSL today for corporate networks and T1 service. Commercial ADSL deployment by most telephone companies began by 1998. Actual deliverable capacity, especially for ADSL, depends heavily on loop conditions.
Cable modems	1.2–27 Mbps (shared capacity)	128 kbps–10 Mbps (shared capacity) or POTS line used for upstream	Several companies are deploying infrastructure (e.g. @Home, Comcast, Time-Warner). Many technical and economic questions remain.
Wireless	28.8 kbps (900 Mhz) 1.5 Mbps (LMDS) 1.5 Mbps (MMDS)	28.8 kbps (900 Mhz) 1.5 Mbps (LMDS) 1.5 Mbps (MMDS)	These are only some of the technologies under development that could provide wireless Internet access (NII/Superset band and 2.3 Ghz auction may also open spectrum for this application). Actual capacity will depend on environmental factors as well as details of deployment.
Satellite	400 kbps (DirectPC)	POTS line used for upstream	Several other systems under development.

Source: Werbach (1997), figure 11.

Ciciora's "bits-are-bits" paradigm, however, contains a fallacy. Bits may not be comparable if the circumstances of their delivery are different. The nature of the network cannot be ignored. Television is one-way; telephones are two-way, switched. The second is more functional, and far more costly, than the first. Even if television pricing reflected spectrum usage, television bits would be cheaper than telephone bits. Also, of course, there is far more redundancy in a conventional TV broadcast than in a telephone signal, which lowers the relative value of the broadband bits. In a free market there would be no reason to expect these prices to equalize. Bits may be bits, but communication services are more than bits.

Hertz and Cents

Bandwidth is one key to providing attractive communication services, and cost per bit is the key to competitive commercial success, holding constant the form and content of the delivered images. This does not mean that bandwidth is sufficient. As noted earlier, in an important sense bandwidth is no scarcer than equipment, storage, or other substitutable ingredients of communication. Moreover, a television channel has 600 times the bandwidth of an AM radio channel, but it is easy to imagine programming on the TV channel that drives away most of its audience, and programming on the AM channel that produces a vast audience. Bits are not the same as meaning or value. But assuming that each service provider is making the most effective use of the bandwidth under its control, more bandwidth produces more revenue opportunities than less bandwidth. As a rule, the cost per bit declines as the bandwidth increases.

Conclusion

Network structure affects not only the cost of communicating but also the kind of communication that can take place. Switched networks conserve on transmission pipelines, but are useful only if not everyone is sending or receiving communications at once.

Telephone networks work by opening a physical connection or channel between the parties. This channel is dedicated to their use for the duration of the call. In contrast, a packet network such as the

Internet does not have dedicated connections. All messages are coded into packets of bits, and each packet finds its way through the network by sharing trunk capacity with other packets. Packet networks transmit the same amount of data far more efficiently than telephone networks. Unlike telephone networks, however, packet networks cannot guarantee that messages will arrive without delay or in the correct sequence. When a packet network experiences heavy use, significant delay or "latency" can occur.

Telephone networks have switches, trunks, and local connections or "loops." The bulk of the capital investment in the telephone system is in copper wires running between telephones and local switches.

The practical capacities of trunk pipelines, for example, optical fibers, has been growing rapidly, and many such facilities are in use. Increasing the capacity of the local loop requires either that expensive processing equipment be put at its ends, or that it be replaced by a coaxial cable or optical fiber. Each of these alternatives is enormously expensive if applied to all telephone lines.

Among the more popular current approaches to expanding the capacity of local loops is ADSL (asymmetric digital subscriber line). This and other digital local loop technologies may permit telephone wires to carry television signals. Because of the structure of the telephone network, consumers would then have access to interactive video, including video entertainment on demand and video telephone service. As with equivalent services on cable television, no one knows whether there is enough consumer demand to support the required investment.

Pipeline economics alone are not sufficient to determine the outcome of the competition to become the next successful electronic medium. What consumers purchase is not pipelines but communication services and entertainment products. The form and character of these products and services are determined not merely by the pipeline and its costs, but also by the costs and capabilities of the associated network and by the cost and ease of use of the associated consumer electronic equipment.

CHAPTER ELEVEN

THE INTERNET

The Internet is both a collection of software and hardware technologies and a collection of services. In each respect it has become cheaper, faster, and more attractive. No longer is it only the toy of graduate students, engineers, and nerdy teenagers. Nevertheless, using the Internet is far from being as cheap, easy, and reliable as watching TV. In this book we are concerned with consumers' use of the Internet for video information and entertainment, and advertisers' use of the Internet as a marketing tool, relative in both cases to conventional broadcasting and other technologies.

Even today only a small minority of consumer households have access to the Internet. But a constellation of commercial interests hopes to profit from Internet development. As a result, what people know about the Internet is mostly from hype rather than from experience. Both the commercial part of the Internet and the volunteer sector have obvious network externalities that encourage them to engage in Internet hype. Suppliers of Internet-related hardware and software have even better reasons for hype, as do academics and other experts.

So effective is the hype that a lack of understanding and experience does not prevent politicians and others from urging the universal availability of Internet access, subsidized if necessary at the expense of other communication services. It is interesting in this respect to note that President Clinton, whose government promoted development of the national information infrastructure—the NII, or so-called information superhighway—reportedly did not know how to send or receive e-mail, the most elementary use of the Internet (*New Republic,* Feb. 17, 1997).

Leaving the hype aside, what is the reality? The Internet is a bit like the *Encyclopaedia Britannica,* which has a reputation for being hard to use. It is a great resource, and something one would like to have, especially for one's children. But for most consumers, actual use of it is so difficult and slow that by the time they find the answer, they have forgotten or lost interest in the question. Unlike the *Encyclopaedia Britannica,* however, the Internet thus far has little depth; it has been a mile wide and an inch deep. The windows offered by thousands of home pages on the World Wide Web are enticing storefronts, suggesting great riches inside. But the shelves are often empty or at least disappointing, especially in relation to the time and effort required to see them. All this is changing; but will it change enough to make the Internet into a mass medium for the delivery of television and similar material?

This chapter is devoted to a discussion of Internet technology and services insofar as they are useful in understanding the issue of television and Internet convergence. A comprehensive assessment of the Internet is not attempted.

At its most basic, the Internet can be characterized as an interactive (two-way) communication network linking users with one another and with distant computers (servers). The most fundamental use of the Internet is the transmission of electronic mail, or e-mail. Almost equally basic is the ability of the Internet to link users with remote databases or remote computers, which then come under the user's control.

The transmission of e-mail is like sending letters through the postal service—except faster. Unlike a telephone call, the delivery of a letter opens no dedicated channel through the postal system; mail trucks, sorting stations, and delivery vans are shared by all messages, and so it is with e-mail. Internet access to a database is like calling a reference librarian, asking a question, and waiting for an answer—except faster. Using the Internet to control a remote computer is like using a garage door opener. The remote control signal is evaluated by a microprocessor (a miniature computer), which activates the motor that opens and closes the door, and which stands ready to reverse either action in response to additional signals or sensor readings.

All three basic Internet functions are highly interactive and require two-way communication among users. In this respect, the Internet is at an opposite pole from broadcasting, whether electronic or print. Broadcasting is not interactive except in the primitive sense that the consumer selects a particular TV channel or a particular magazine. That broadcasting is not interactive is a virtue because it permits the use of less costly communication media and equipment. It may also be a virtue in the sense that consumers value passive media consumption. The Internet satisfies entirely different needs with different media.

It All Started with Castro

The Internet started out in the early 1960s as a Defense Department research project. The Cuban missile crisis and the Bay of Pigs invasion had demonstrated grave weaknesses in military communication systems. These were the days before the Vietnam protests and student riots against ROTC; the political climate was favorable for military research on campus. Computer scientists at major universities received sizable Defense Department grants. One aim of the research was to develop communication networks that would be less vulnerable to enemy attack. An experimental computer network called the ARPAnet (Advanced Research Projects Agency Network) was one result, based on the new technology of digital packet switching. Interestingly, ARPA during this period was headed by J. C. R. Licklider, one of the prophets of the information revolution.

In an ordinary telephone network, everyone is linked to a local switch, and the local switches are connected to regional switches, which are linked to each other (see Figure 10.2). If one of the switches in this network is destroyed, users connected directly or indirectly to the switch lose some or all service. (In practice, the United States telephone system is not this vulnerable, because there are crosslinks among the switches, providing alternate paths through the network.)

A digital packet network has several advantages over an ordinary analog telephone network. First, in an analog system, an open line or "connection" between two users chews up a whole line's worth of capacity, whether or not anyone is talking. Silence, even a small pause, wastes bandwidth. Second, a standard telephone system relies on a star network, which means that congestion on any particular link or at any particular switch can prevent a call from going through. In contrast, a digital packet system transforms any message (data, voice, video) into a digital stream, and then breaks down the stream of digits into packets, or bundles of bits.

Each packet contains not only the information content of the message but also the addresses of the sender and the recipient and other "overhead" data. Each packet is sent off to a network "router," which forwards the packet to another available router along what it calculates to be the fastest path to the intended recipient. In principle, each packet of a given message can take a different path through the network, and packets may arrive out of order, late, or not at all. At the other end, the recipient's computer reassembles the message. The receiving computer acknowledges receipt of each packet; failing such acknowledgment, the originating computer repeatedly resends the packet.

There is a tradeoff between computing power (at the router and at the two ends) and transmission bandwidth. A packet network fits far more data into a given bandwidth than does a standard telephone connection because the packet network makes no connections; therefore, idle bandwidth is minimized. Further, a packet network maximizes reliability at the expense of output. Packets arriving at a busy router are dropped. They must be sent again and again until the destination computer acknowledges receipt. Latency, the unpredictable delay before a given packet reaches its destination, is a serious problem for "real-time" and interactive transmissions.

A packet network can survive the destruction of one or more of its routers and, because it squeezes every ounce of capacity out of the communication lines connecting the routers, it can handle much more traffic than an analog network with the same physical facilities. In one sense, this network architecture is an example of the phenomenon of digital compression, a driving force of the information revolution.

When military funding ran out and the National Science Foundation took over, the ARPAnet became the NSFnet, whereupon it began to grow rapidly under the control of academics. Through ad hoc committees, debates, and experiments, they developed standards and protocols ensuring connectivity among campus computer networks. Most significant was the development of the Internet Protocol in 1982. The IP standard specifies the form that packets may take and establishes the rules for how routers will deal with them. Soon after came the Transport Control Protocol (TCP), which governs error correction and causes the receiver's computer to ask the sender's to resend data until the message has been correctly received. Basic applications followed—Telnet, Gopher, Archie, WAIS, and finally the World Wide Web in 1993. Commercial research organizations were drawn in, both because the content of the Internet was heavily research oriented and because the e-mail capabilities of the Internet facilitated communication within the research community. NSFnet was based on the same technical premises as ARPAnet, plus the tacit assumption that data storage costs were too great to justify local storage and maintenance of copies of large databases. If this assumption does not remain valid, storage capacity would replace some transmission capacity in order to minimize cost.

Who Pays for the Internet?

The National Science Foundation paid for the Internet "backbone"—the high-speed digital communication links among the routers—after the Defense Department stopped funding the ARPAnet. Telephone companies and other commercial contractors provide the physical facilities. The backbone in 1975 had a speed of 56 kilobits per second, the same as an ISDN channel today, or the fastest analog modem. By 1981 the backbone was upgraded to what telephone engineers call a "T1" channel—1.554 megabits per second. A T1 channel is a great deal of bandwidth by telephone standards, but it is nowhere near enough for conventional television signals. In 1991–1992, as the NSF was phasing out its subsidies, the NSFnet backbone was upgraded once again, to T3, or 45 megabits per second, enough for five to ten compressed TV channels.

The Internet was effectively privatized between 1992 and 1995. Thereafter, communication companies cooperated with one another in providing pieces of the network (transmission facilities, access points, routers), while industry committees and the Commercial Internet Exchange facilitated agreement on interconnection standards. These companies are in business to make money, and they are paid by those who want access to their facilities—chiefly large companies, universities, and Internet service providers (ISPs). Forty or more national ISPs today provide backbone facilities for the Internet as well as access points for smaller national and regional networks, institutions, and users. These ISPs include MCI, Sprint, Alternet, BBN, PSI, and AGIS. There are several thousand smaller ISPs, providing user connection to the Internet but not backbone service.

The ISPs are private, for-profit communication companies. They build or lease facilities to provide transmission and routing services. Their revenue comes from their customers, who pay to connect with and through them. The ISPs compete both in price and in quality of service. Quality of service includes such factors as latency and reliability. The big ISPs also connect with one another and exchange roughly equal volumes of traffic. It is customary for such "peer" networks not to pay one another for accepting the other's traffic, presumably because transaction costs can be avoided in that way with no significant net revenue loss on either side. Small national and regional networks, and foreign networks, are treated more like customers than like peers. Generally, such networks pay a transit charge to the larger ISP for delivery of the messages originating on (or transiting) the smaller network, especially if there is reason to suppose the traffic flows will be asymmetric. For example, a Web site like CNN or Fox News that sends out hundreds or thousands of simultaneous video bitstreams in response to requests from users is highly asymmetric. An ISP or node serving a number of such sites imposes heavy costs on the backbones of other providers.

One feature of the Internet has an interesting parallel in electric power networks. Because of the connectionless nature of the Internet, the path taken by individual packets may wander anywhere in the net. The routers of mortal enemies may be, and probably are, processing each other's packets, especially during times of congestion on the backbone. This can complicate both pricing and security issues. An analogous situation arises in high-voltage electricity networks that carry wholesale power from one utility to another. Because the electrical transmission lines are interconnected, there is no preventing some of the electrons from taking a path other than the shortest one from buyer to seller, much less the path on which buyer or seller may own or have purchased transport capacity. (The phenomenon is called parallel path flow.) Some economists believe that effective competition in wholesale electricity transport is impossible for this reason. So far, however, it does not appear to be a problem for the Internet.

As price and quality competition has heated up among communication companies, the concept of a single Internet backbone has disappeared. Although the large Internet backbones in the United States are not isolated from one another (or else they would not be part of the Internet), they are nevertheless able to maintain distinct service quality standards. Technicians can isolate service quality issues arising in the serving backbone from those arising in servers, local connections, and other components. More ISP backbones are under construction. In this respect the Internet is much like the competitive long-distance telephone industry; interconnection does not impede quality and price competition. The analogy goes further, for while Internet and switched telephone long-distance facilities generally are distinct, they come together, for now, on the last mile of their journey to the user. The wires on that mile are owned by local telephone companies for whom, as yet, there is little or no competition.

Are Web Surfers Still Nerds?

Statistics on Internet use are not yet reliable. One major source is a semiannual survey conducted by researchers at Georgia Tech, but their sample is self-selected and thus not representative of any but (perhaps) the most enthusiastic surfers. Nielsen conducts a broader survey intended to be representative of the entire population of persons with telephones. However, the Nielsen survey is sponsored by CommerceNet, an organization dedicated to promoting commercial exploitation of the Web. Other surveys also have a background perspective that casts doubt on their conclusions. Eventually, reliable statistics will become available because advertisers and other commercial users will demand them.

Unrepresentative and nonrandom though it may be, the periodic survey conducted by the Graphic, Visualization, and Usability Center (GVU) at Georgia Tech University paints an interesting picture of the active Internet user over time. The early stereotypical user was a male engineering or computer science undergraduate with a weakness for pornography and a strong interest in "hacking," or making unauthorized entry into distant computers and networks. The current user of the Internet is quite different, according to the November 1997 survey of Internet users active enough to have seen Georgia Tech's invitations and interested enough to respond.

The mean average household income of responding Internet users in the United States was $53,000. This figure has declined over time; when Georgia Tech first began collecting household income data, in May 1995, average household income of U.S. Internet users was $69,000. Nonetheless, Internet users tend to have much higher incomes than the general population. In 1996, the U.S. median household income was $32,264.

One clear trend identified by the Georgia Tech survey reveals a narrowing but not yet convergent gender gap in Internet usage. In 1994, the first year of the survey, an astonishing 95 percent of responding Internet users were male. By 1995, the percentage of male users had dropped to 82 percent. By 1996, it had been reduced to 69 percent, and by 1997 to 62 percent. Though a large increase in the number of female users occurred in 1995 and 1996, the gender ratio has remained consistent in recent years. The increasing use of the Internet by college students probably facilitated the rise in the number of women connected to the Internet. But a study of proficiency in Web browsing, reported in the press in the summer of 1997, found that older women (over fifty) were more proficient than younger men or women.

As has been the case since the World Wide Web became available, users of the Internet remain a highly educated group. Forty-seven percent of respondents have completed a college or advanced degree. That figure has remained virtually unchanged since 1995, when 56 percent of Internet users reported having attained a college diploma. During the past three surveys, a higher percentage of women have reported some college experience, lending support to the notion that Internet use at colleges and universities has narrowed the Web's gender gap.

The Georgia Tech survey also hints at the occupations of Internet users. Though users of the Internet have become much more diverse since 1994, it remains a medium dominated by computer- and education-related professionals. As of late 1997, 21 percent of responding Internet users reported being in a computer-related field and 23 percent reported working in education. Another 21 percent listed themselves as professionals, and 11 percent said they were in management.

An important consequence of the development of the Internet and the World Wide Web is that users inevitably must sacrifice other leisure activities to have time to "surf the net." The activities foregone presage the role that the Internet will play in consumers' lives. A study by the media research firm Find/SVP was sponsored by thirty major companies in the computer, networking, and online service industries. Respondents, one thousand randomly dialed Internet users chosen from focus groups and self-selected online surveys, were asked which other media were affected by their Internet use (either positively or negatively). By far the most common answer was that television viewing decreased. Among those who responded that television viewing had declined, the majority were people whose Internet use is 50 percent or more for personal reasons, especially users connected to the Internet through America Online. Assuming that the survey results are credible, we see here a strong hint that Internet use for some consumers may be a substitute for television viewing. The November 1997 Georgia Tech study also asked Internet users how often they sacrificed television to use the World Wide Web or send e-mail. Close to a majority of responding Internet users claimed to watch significantly less television, choosing instead to surf the net several times a week or even several times a day. Finally, America Online announced in September 1997 that 37 percent of its subscribers reported less TV viewing than in the past.

Because television viewing is often a secondary activity (in the background while eating, chatting, reading, or sleeping), it is possible that Web surfing, with its long delays and lengthy downloads, may encourage secondary TV viewing as well. The answer depends in part on whether the computer and the TV are in the same room. According to a survey conducted in 1997 by NPD, 40 percent of all PC households watch television while using the computer.

What's It Good for?

Today the Internet is a medium for many different services, most of them mediated by the ubiquitous Web browser. New services appear hourly, and older ones disappear. The transition from the academic tradition of free service to commercial service is nearly complete. In 1996, two-thirds of Internet users polled by Georgia State researchers approved of fees for Internet site visits, an idea that would have caused a revolt a few years ago. To begin to understand where the Internet may be headed, it is useful to survey its past.

E-mail

The Internet was originally a means by which military communications (electronic mail) could be reliably exchanged under challenging conditions. It was designed by academics who delighted in the ability to enhance collaborative research by exchanging their own e-mail. From there the e-mail habit spread to students and then to many outside the academic world.

CompuServe offered e-mail to specialists (those subscribing to CompuServe) beginning in 1979, but only recently have ordinary citizens been able to use e-mail for private communications. A major facilitating factor is the Simple Mail Transfer Protocol (SMPT), which permits computers of all kinds to exchange e-mail messages over the Internet.

Like fax machines and other services with network externalities, e-mail has begun to grow very rapidly as a critical mass of consumers begins to use it. E-mail is useful, of course, only if the people with whom one wishes to correspond can receive messages and are in the habit of checking their mailboxes. Since only about 40 percent of all American homes have computers, and less than half of them have modems or are connected to ISPs or to services such as America Online, we have perhaps not yet reached critical mass at the household level. But on the other hand, many people probably use e-mail facilities at their workplaces for personal communication with correspondents outside the office. Further, e-mail penetration among the youthful professional classes is probably extremely high—so those whom an e-mail–equipped person is likely to want to reach are likely to be reachable.

For those who use e-mail, it has a great advantage over both the telephone and the postal service (a.k.a. "snail mail"): it is free. (Compared with the postal service, e-mail is, of course, also speedier and more convenient.) Apparently many people who pay $20 per month to AOL or other access providers do so chiefly to send and receive e-mail. Whether this is cheaper than the telephone depends on usage; with overseas friends and relatives there is a great saving over long-distance rates. The disadvantage of e-mail, relative to the telephone and the postal service, is that many people, even those with access, do not use it and are therefore unreachable. The usefulness of e-mail is directly proportional to the number of others who use it, because of network externalities. Of course, e-mail is less interactive than the telephone, and perhaps less flexible than the postal service in terms of the forms messages may take. (In order to convey the idea of emphasis in an e-mail message, one must resort at present to devices such as *emphasis*, and although formatting options will improve under the new SMPT protocols, it is still impossible to send a perfumed note.) Still, it is clear that e-mail has been and will remain one of the principal sources of consumer value from the Internet.

Telnet

After e-mail, Telnet is the oldest widespread Internet application. Telnet permits a distant user (originally using a "dumb" computer terminal rather than a PC) to log into a compliant computer and search its files. Many library card catalogs, including that of the Library of Congress, were put on Telnet. Telnet was useful for information searches, but aside from the information in the directories and indexes, there was no way to transfer data from the Telnet host computer to the distant user. In other words, one could search the card catalog but not the books themselves. Interestingly, the Library of Congress is now working on a project to digitize its entire collection, starting with the oldest works that are in deteriorating condition. Eventually, all of the information in the library (that is not subject to copyright protection) will be accessible on the Internet—but not through Telnet.

Me and My Database

Almost as old as e-mail and Telnet in the history of the Internet is the idea of sending "computer files"—that is, a body of data—from one user to another. In order to do this, users must agree in advance on certain protocols and standards, so that the data stored on user 1's computer will be meaningful to user 2's computer. The system that was developed for this purpose is called the File Transfer Protocol (FTP). Using applications that implement FTP, user 1 can maintain a set of files on a computer (called a "server"); these files are accessible to any other user connected to the Internet and can be "downloaded" (copied) to any user's computer. FTP was originally designed to permit computer scientists to exchange research data. But FTP has been applied to the transfer of all kinds of data: word-processing documents, digitized maps, photographs and other images, computer programs, databases, audio recordings, compressed video recordings, and so on. Any discrete file in digital form can be transferred over the Internet using FTP.

As the number of FTP servers grew with the spread of the Internet, it became more difficult to say where any particular file was located. One might suspect a useful file existed but not know where to find it. Gophers and related applications (Archie, WAIS) were developed to deal with this problem. Persons in charge of FTP file servers would register in their directories not only the name and Internet address of each of their files, but also descriptive information such as key words. Gopher and like programs would rapidly search the entire Internet for key words, authors' names, or other identifiers specified by any researcher. Thus it became possible to search the Internet for information that one did not know existed—and find it. Searches became an increasingly powerful and necessary tool as the Internet grew.

Of course, backbone bandwidth must increase as the number of users and usage increases. In the late 1970s, when every major university in the United States and many faculty and students were connected to the Internet, the shared backbone communication links had a bandwidth or speed of only 56 kilobits per second. This same speed is now regarded by many home computer users as barely adequate for their own private connections to the Internet. This is because we have moved on from e-mail, FTP, and Gopher, all of which involve chiefly textual material, to the World Wide Web, which makes intensive use of graphical materials.

Smutnet

Although there appear to be no objective measurements of it, pornography has been one of the most important motivating forces for Internet use, just as it was in the early development of videocassette recorders. Downloading a file that can be reconstructed into a pornographic image may seem even more private to users than viewing a videocassette purchased by mail. (In neither case is the user's identity a secret to the supplier, but at least there is no intervening postal worker.) Especially in the early days of the Internet, downloading an image such as a digitized page from *Playboy* could take hours. Still, this activity, strictly amateur at first but today highly commercialized, seems to have been responsible for much of the early demand for Internet access, just as it was responsible for the early use of VCRs.

Even today, sex is practically the only clearly successful commercial use of the Internet. Internet sales of sex-related merchandise and online services are booming ("As Other Internet Ventures Fail, Sex Sites Are Raking in Millions," *Wall Street Journal,* May 20, 1997).

Keeping in Touch: Usenet and Bitnet

Although now interconnected with the Internet, Usenet and Bitnet were once distinct networks with similar features. Both are designed to permit individuals anywhere in the Internet world to share and exchange views and information centered on a common theme, hobby, or interest. Usenet was developed in the late 1970s as an application for interconnected mainframes running the Unix operating system, a favorite among academics. Each mainframe or host system maintains a set of files called "newsgroups," each corresponding to what in other contexts are called chat rooms, bulletin boards, conferences, and special interest groups. In 1994 there were over 5,000 topically defined newsgroups. As of March 15, 1997, there were 8,663 organized enough to have an anchorman or newsreader, and over 9,000 in February 1998. Anyone with access to a local host can read and participate in newsgroups, which are like an extended multiparty conversation, devoted at least in principle to a narrow substantive topic ("Elvis sightings").

Usenet postings are circulated by means of interconnected neighboring mainframes or servers updating one another's files. New postings find their way around the world, from one host to another. Whereas the initial Usenet hosts were university mainframes, nowadays the hosts are ISP servers. Because of newsgroups' often controversial content, not all online and ISP services provide access to them.

Bitnet is an international network that operates by means of e-mail but accomplishes much the same purpose as Usenet. Bitnet host computers have software called "listservers." Anyone interested in the topic covered by the server signs up and in return receives, via e-mail, automatic periodic updates of postings on the server. Postings are sent in by e-mail as well. Some listservers are entirely automatic: members of a Usenet newsgroup, for example, automatically receive everything on a topic. Other listservers are edited or moderated.

Bitnet and Usenet activities are highly interactive, like e-mail, but listservers, especially, also contain an element of broadcasting. Each message is "one to all." In this sense messages are like letters to the editor in the local newspaper. So-called chat rooms on services such as AOL have a counterpart on the Internet—Internet Relay Chat (IRC), which works in a similar way.

The World Wide Web

Before the invention of the World Wide Web/hyperlink concept in 1990 by Tim Berners-Lee, a researcher at CERN, the European high-energy physics laboratory in Switzerland, and the subsequent (1993) development of the Mosaic Web browser at the National Center for Supercomputer Applications (NCSA) in Illinois, the Internet consisted of users with personal computers or mainframe terminals, all connected to one another and to file servers and Telnet sites. After the invention of the Web, the Internet consisted of exactly the same components and performed essentially the same functions as before. But instead of relying on the clumsy and rigid applications available previously, each user, or each user's computer, could turn a colorful, iconized face toward the Internet, a face that made it far easier for users to search for and to convey information. In this sense the World Wide Web is nothing more than the Internet with a "friendly face."

The World Wide Web and Web browser, though astoundingly successful, were not the result of complicated high-tech research. Illustrating this point, Microsoft's chief technology officer, Nathan Myhrvold, had the following reaction to the revolutionary Mosaic Web browser:

> If you'd said up front, "My research program is that I'm going to allow bit maps to get transferred over this simple protocol" [the function of a Web browser], people would have said, "That isn't research." It isn't! It turned out that a low-tech social phenomenon called the Internet has suddenly arisen and surprised people. But it's like asking people in plastics research why the hula hoop was successful. (quoted in *Upside,* Internet publication, April 1997)

The key to the success of the Web is the ease with which it permits associative searching, a concept first put forward by Vannevar Bush (1945, 101–108) in a famous *Atlantic Monthly* article. Users seeking data on the Web no longer need to search laboriously in FTP directories with arcane and unhelpful names, using rigidly formatted commands designed for machinery, not humans. In its simplest form, the Web overlays such directories with a graphical interface. Objects or images or words on the interface represent files (or data in various other forms, such as instructions to the user's processor). A user may select such an icon and the interface will recognize the selection as a command to produce (transfer) the indicated information across the Internet to the user. The "indicated information" is preprogrammed by the author of the Web page. Such procedures are called "hyperlinks." Thus simply using a mouse to click a "button" or other object on the screen often produces very powerful results, such as the immediate transfer of megabytes of data from one computer to another, or the initiation of a series of processor instructions, such as instructions to retrieve and display data from a user's hard drive or CD-ROM disk.

"Hypertext" is a specialized application of hyperlink technology. Hypertext appears on the screen as ordinary text, but if selected (clicked) replaces the current screen with some other screen as determined by instructions found at an address on the Internet (or locally) that has been, in effect, hidden behind the original text. With hyperlinks, text is now icon. According to Vannevar Bush, research (in the broadest sense) supported by such links keeps the mind free to roam the world's knowledge, unhindered by the abstract and irrelevant structures and categories of library card catalogs and file directories. The effectiveness of the technique, of course, also depends on the artistry of the author of the Web page. Still, in principle one can click "associatively" through a significant fraction of the world's digitized information, going from one Web page to another, link by link.

For the World Wide Web to operate as it does, the computers at both ends of each transmission must use the same "language." This permits a distant server to project content, some of which may be functional, on each user's screen. The "projected interface," which appears to be a window into the inside workings of someone else's server, is a manipulation of the local interface by distant instructions. In the language of the Web, servers are "sites" with "home pages." Users contact sites by using a "browser"—software designed to make the user's screen take on the appearance commanded by whatever distant site is being contacted.

The central feature of the Web that distinguishes it from the Internet as it was before is ease of use. Web browsers and even applications for constructing Web sites are intuitive and accessible to many nontechnical users. But Web browsers and Web sites still do nothing that could not be done, more slowly and with greater difficulty, with e-mail, FTP, Telnet, Gopher, WAIS, and Archie.

What You Can Do with the World Wide Web

A Web "site" is a server—often simply a personal computer with a hard disk drive and a modem—connected to the Internet by a phone line. More commonly, nowadays, commercial services provide host servers for hundreds or thousands of individual Web sites. (Such a host consists of a more powerful computer attached to multiple phone lines, and the ability to operate many Web sites simultaneously.) In either case, each server has a unique address that has nothing to do with its physical location: www.whitehouse.gov, for example, or www.nytimes.com, either of which could be in Kansas.

A server has data files that can be transmitted to the computer of the inquiring user, upon the user's selection of a hyperlink command. Communication is mediated by the Web browser application. An initial inquiry produces an image on the user's screen of the "home page" of the distant server. A home page has text and graphics, much of which conceals directions to other data on the same or other servers. A mouse click invokes these directions and induces the transmission of more data: perhaps another page, a file, or an audio or video clip.

One can think of Web sites or Web home pages as "channels" in the television sense. They are programmed with content intended to attract viewers; they have advertising; some charge a fee for admission. Audio and video components are increasingly common. Many servers do not cease to transmit data once a complete screenful has been sent, but continuously update some portions of the screen. Some emit "streaming" (continuous) audio or video as well as graphical information. Others simulate the effect of continuous updating by inducing movement created locally by the viewer's computer, acting on instructions from the distant server. (For example, a server can instruct the user's computer to flash a message repeatedly on the screen, producing the same effect as if the server were sending the same message many times in succession. This is another instance of computer-bandwidth tradeoff.) Pages that are completely passive—just text or images, no hyperlinks or streaming data—seem boring by comparison. Of course, heavy use of graphics and animation with redundant content can actually reduce the amount of information presented by a Web page, making it more like television and less like a newspaper or book. Ironically, it takes more bits to transmit less information, catering to the tastes of passive users.

The substantive content of Web sites varies enormously. In terms of numbers, most are the "personal" pages of individuals, typically high school and college students. But the lure of commercial profits has attracted thousands of retail establishments who hope to advertise their existence and goods, and to make direct sales over the Web. There are few large corporations that do not have a Web site devoted to public relations goals. Mail-order houses, especially those selling computer hardware and software, books, and music, are found on the Web in abundance. As noted, a large sector (though probably a declining proportion) of the Web is devoted to pornography.

Finally, many traditional media have developed Web sites. Among the more elaborate and successful to date are those of the *San Jose Mercury-News,* the *Washington Post,* the *New York Times,* and the *Wall Street Journal.* Some newspaper sites offer content that is quite distinct from the published version. Some charge for access to portions *(San Jose Mercury-News)* or all *(Wall Street Journal)* of their site; others *(Washington Post)* do not. Broadcast television and radio media also have developed elaborate, though in many cases still experimental, Web sites. CNN and Fox News both send out "streaming" audio and video data, which permit the user to view very fuzzy, jerky video images in a tiny window on the computer screen. The rest of the page or screen consists of text and icons directing the user to other portions of the site, but typically not to information related to what is happening in the video window.

Two Grains of Wheat in Two Bushels of Chaff

Associative searching, wandering the Web from page to page in search of that certain detail, is literally an endless task. New pages and new data are added much more quickly than a user can scan them. Help is needed, and search engines have come to the rescue. Search engines are computers whose "back ends" constantly wander the Web, though very rapidly, gathering addresses and indexing content. The "front end" of the search engine is a user interface (a Web site) that any Internet user may command to search the current back-end index for specific words and phrases. The result is displayed as a list of hypertext entries, giving information such as the name of the site (if any), the title of the specific document (if any), its date, brief excerpts of text, and so on. The user can click on any of these and be sent to that site.

Because the Internet (and newsgroups, which are also searched and indexed) is very large, searching for specific information can be daunting. A single example may suffice: one popular search engine, Alta Vista, in March 1997 found about 300,000 Web pages and about 4,000 Usenet newsgroup documents that mention the word "economics"; 500,000 Web pages and 20,000 newsgroups documents that mention "television"; and 30,000 Web pages and 200 newsgroup documents that mention both "television" and "economics." By March

1998 the totals were "economics," 676,000 Web pages and 20,200 newsgroups; "television," 1,381,000 Web pages and 23,200 newsgroups; and for both terms, 48,000 Web pages and 401 newsgroups. The longer I wait to do the research for this book, the smaller the fraction of available "knowledge" I can hope to review.

Search engines are essential tools in making effective use of the Internet. Nevertheless, search engines are able to cover only a minority of Web content and remain wholly text-oriented. No one has yet developed a similarly useful tool for searching the nontextual information that can be represented directly on the Internet, such as graphic images and video streams, much less information that can be represented only indirectly on the Internet, such as emotions or odors. (Emotions and odors are information, despite our limited ability to transmit them.)

Search engines compete by offering differentiated products; each is constructed somewhat differently and has different strengths and weaknesses. They also differ in responsiveness. Search engines are by far the most-visited sites on the Internet and in consequence make excellent advertising vehicles. Search engines are now the chief recipients of advertising revenue on the Web, surpassing AOL.

The Front Door: Internet Service Providers

There is a spectrum of Internet service providers or online services, from those that do little more than sell access to the Internet to those that provide additional proprietary ("value added") information services (airline reservations, stock quotes and transactions, and electronic banking, for example). The best known of the latter are America Online (AOL) and CompuServe, now owned by AOL. Both came into existence before the Web and the Mosaic browser were invented, and hence before access to the Internet was of interest to most nonacademics. What these online services provide today is access not only to the Internet but to a constellation of information services very like the Internet, and a standard user interface. Some of the information is published by the online services; but most information is provided by outside organizations and by the users (uploaded shareware software, for example). The online services also supply e-mail access and "bulletin boards" or "chat rooms" where individuals can engage in extended conversations, in real time or not; these greatly resemble the older Usenet newsgroups and Internet Relay Chat.

As the Web and the Web browser exploded in popularity in the mid-1990s, much of the growth came at the expense of the online services. In contrast to AOL and CompuServe, the Web was more relaxed, more fun, and best of all, free, at least to those affiliated with or employed by an organization that paid for access. Also, Mosaic, Netscape, and MS Explorer provided better user interfaces than those provided at the time by the online services. What the online services did do that independent Web sites could not do at first was to provide an apparently secure framework for commercial transactions. One assumed, if one purchased something through an AOL service, that AOL had screened the service for reliability, or at least that the seller had a stake in good performance in order to maintain access to the online customer base. Similarly, the seller knew something about the provenance of the buyer. The risk and uncertainty associated with transactions are reduced by this vetting of participants. Mechanisms to accomplish this form of security have now been developed on the Web, and therefore, AOL and its ilk no longer stand out as islands of relative security in a chaotic digital world.

After the NSFnet was privatized, the job of providing (1) backbone links and (2) interconnection services between individuals and businesses and the Internet fell to Internet service providers. A pure ISP supplies only interconnection service, although several also provide nationwide Internet backbone services. Users call up the ISP with a computer modem, and an ISP modem answers and connects the call to an Internet router. After independent, bare-bones ISPs pioneered the market, and after America Online and CompuServe began to lose customers, the value-added online services themselves began to offer Internet access service to their subscribers, and even to permit nonsubscribers access to certain portions of the service via the Internet. This trend has continued so that AOL's content is now available "free" to users arriving from the Internet, and many AOL subscribers rely on AOL chiefly for access to the Internet. As a result AOL, like other commercial suppliers on the Internet, has been driven increasingly to rely on advertising revenue.

AOL's (or any ISP's) "point of presence" in a given city in effect consists of a room full of modems, each answering incoming phone calls from users. AOL has several hundred thousand such modems around the country. Each modem is connected by a high-speed communication link to the central AOL computer in Dulles, Virginia, where users interconnect with AOL's various proprietary services. Among these is a router that provides Internet access. The distinguishing feature of the proprietary services is that AOL charges users for them, and that ISP or access services are bundled with content services. (However, AOL offers content-only service to users arriving via the Internet.) Increasingly, individual Web pages are proprietary in this sense and may also serve as gateways to portfolios of proprietary services. Similarly, large ISPs such as AT&T are now offering both content and advertising. It is therefore more and more difficult to distinguish services such as CompuServe or AOL from ISPs or commercial Web sites in general.

Founded in 1969, CompuServe is nearly as old as the ARPAnet. Until rather recently, it did not think of itself as an ISP. Although university faculty and students could reach CompuServe over the Internet in the days before the World Wide Web, they usually used a modem and a dial-up connection. In CompuServe's early days, modems chugged along at 300 bits per second, an unacceptable speed today and excruciatingly slow even then. Modems that transferred data at 1,200 bits per second did not arrive until the mid-1980s.

CompuServe and the other online services have had a difficult time adjusting to the Web. They are faced with the need for enormous investments in communication capacity (to accommodate the consequences of flat-rate pricing) at the very time that competition from ISPs has caused prices to plummet. CompuServe in 1997 announced plans to abandon the consumer market on account of lack of profitability; shortly thereafter it merged with AOL. The others, including America Online, struggle to deal with the capital investment consequences of the flat-rate pricing to which competition and the lure of advertising dollars have driven them. Most online services appear to believe that their salvation will lie in advertising revenue, which of course should increase as flat-rate pricing increases usage time.

But a more fundamental shift is in order. Right now the online services see themselves as offering access to a shopping mall of information and entertainment services, one of which is Internet access. In an alternative model they might act simply as a shopping mall on the

Internet, providing as a unique value-added service user and vendor vetting to facilitate commercial transactions. The online services could then abandon their consumer access subscribers (though not necessarily simple ISP service) and limit themselves to serving visitors from the Internet and retailers, on a pay-per-visit or -per-transaction basis. It will be interesting to see if AOL manages to survive this transition.

The Medium Is One Message

The physical nature of any transmission medium affects the form and content of the information. For example, the television viewing experience is quite different today, with electronic tuners and remote controls, than it was in the 1960s, when changing the channel meant trekking across the room to the TV set. From the point of view of consumers, the outstanding physical characteristics of the Internet medium are its speed (or lack of it) and its interactive nature.

An ordinary TV set is interactive in the sense that it reacts to users' commands to change stations. But once selected, it is the station and not the viewer that decides the content and the pace at which data flow to the viewer. This is as if ordinary books came equipped with automatic page-turners or scrollers that gave the reader no opportunity to vary the speed at which the material appeared, an analogy that comes home most strongly when one recalls trying to read fast-moving credits at the end of a TV program or movie.

The Internet is interactive in a more fundamental way than television: the user not only chooses each screenful of images and words, but also decides how long that screen will stay active. To operate a Web browser, one must fiddle with mice and keyboards—participate in the experience. In most cases one does not sit passively and let a Web page unfold. As Marshall McLuhan would have put it, the Web is a hot medium. To the extent that the Internet becomes a video medium, that may change. But even if it does, the video content is likely to remain interactive, and only part of the picture.

As discussed in Part I, television viewing can be regarded as a rather passive activity, while Internet use is the opposite. One can conclude from this that television and the Internet serve entirely different human needs. However, as David Ellis (1996, 6) notes, "viewers can't all be put neatly into active and passive psychological profiles." Even TV viewing involves a minimal level of activity, he points out. The viewer must turn the set on, change channels, adjust volume and picture. He continues:

> "Active" can also be understood in an intellectual rather than physical sense. We will therefore take "active" to refer to the mechanical storage or mental retention of program content for some instrumental purpose other than passive viewing. Thus, while timeshifting a broadcast program on a VCR is an active function, viewing it from start to finish as one would on TV would qualify as passive. Definitions like this are not cut and dried: for example, is a timeshifted movie viewed without its commercials by zapping with a remote control an active or passive viewing experience? The answer might be: more active than without the zapping, but much less active than playing a video game.

It is useful to think of these levels of activity in terms of a spectrum. At one end are the passive viewers, "who do not want to be faced with complicated controls to achieve any task beyond finding and watching a program." Active viewers, at the other end, use video "for work-related tasks" and want enough controls and functions "to get things done quickly and enhance their productivity."

Ellis goes on to define passive and active entertainment:

> From the programming perspective, passive entertainment is the dominant form of conventional broadcasting. It is one type of "escapist" leisure pursuit, comparable to reading a novel, the point of which is to become engrossed and surrender choices to the creator (sometimes including the choice of turning off the set and going to sleep). . . .
>
> Active entertainment is wide-ranging, comprising both established and emerging technologies. Among the established group would count both the VCR and video games. The emerging technologies constitute a more complex set, partly because they include various . . . platforms . . . which . . . operate as standalone TV peripherals, as well as an expanding group of networked interactive technologies. (Ellis 1996, 6)

The spectrum of activity represented by broadcast television at one extreme and an Internet chat room or a video game at the other is best viewed as a demand-side and supply-side issue. Specialized media cater to consumer needs at various points on this spectrum. Broadcast television is a medium as well as a collection of products. At the opposite extreme lie telephone networks and the Internet. When considering whether all digital media are likely to "converge" into a single pipeline, one must ask whether that pipeline can satisfy all consumer needs, active as well as passive.

On the supply side, conventional television supports a huge program production industry in Hollywood and elsewhere. Production of attractive programming material for the Web is much more heterogeneous in its origins and content, and the majority is not entertainment but data. Owners of Web sites value popularity, whether because they sell advertising, or merchandise, or admission, or simply from pride. The form in which the information is packaged affects the popularity of the "program" (site) and hence is often intended to be entertaining. For example, a search engine such as HotBot, which performs the most prosaic of data-processing tasks, is likely to have colorful 3-D buttons, bouncing icons, and screaming streaming crawlers.

Although certainly a medium in its own right, the Web is closer to print than to video in its form and in its relationship with the user. Reading is a less passive activity than TV viewing; it is more like Web surfing, which is, after all, chiefly reading. It follows that those who are successful at designing attractive printed material are in a better position, initially at least, to design attractive Web pages than those who design video material. Indeed, one major issue that will affect the extent to which Internet and television "converge" is the extent to which video programmers learn to program successfully for the Web.

On the Dole

The National Science Foundation is subsidizing development of a new, much faster network called Internet II or vBNS (for very high-speed backbone network service). It will connect five national supercomputer centers (see Figure 11.1). Other access points will be connected as well. Internet II is designed to permit massively parallel computing, in which a number of coordinated computers tackle the same problem

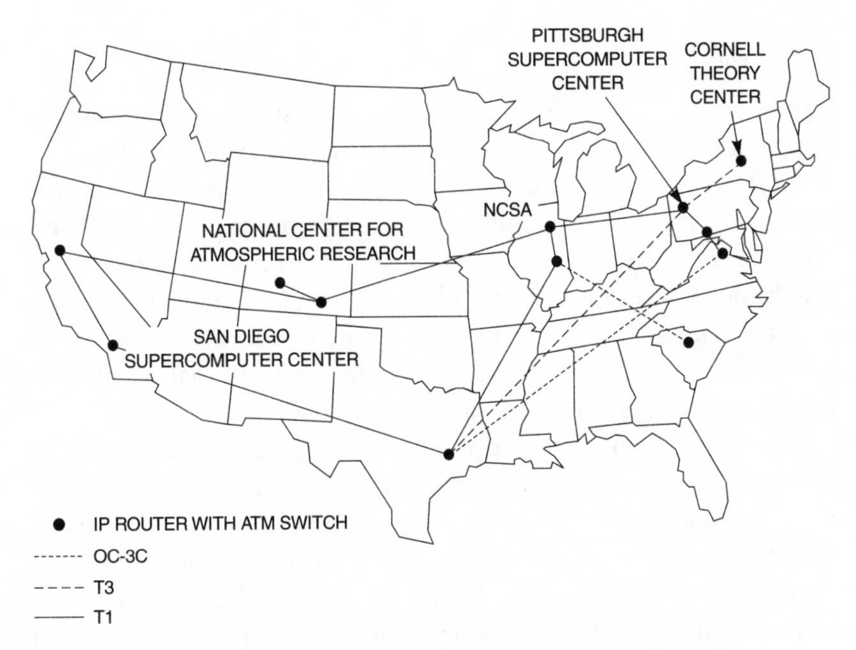

Figure 11.1 The vBNS Internet II (National Science Foundation, 1996)

simultaneously, among other activities. At first this new network will not be part of the original Internet, but it is scheduled to be privatized in 1999, and the two may then be interconnected.

It is undeniable that government funding was crucial to the development of the Internet, and that government funding ended at around the time that the Internet became a commercially viable industry. It is difficult to think of a more successful use of government subsidies, or a more appropriate method of funding such research. The National Science Foundation deserves great credit for exercising the will to terminate subsidies for Internet I despite pressure from powerful academic vested interests.

Unfortunately, the example set by the National Science Foundation may not be followed by the rest of the government, which seems anxious to promote the use of the Internet. Promotional efforts take the form of mandated cross-subsidies, in which users of other telecommunications services may be asked to pay some of the costs of some or all Internet users.

I Play, You Pay

The Internet is a network of networks. At first, this meant that the professors at Stanford, connected to one another by a local area network (LAN) or by links to a local mainframe computer, and the professors at MIT, similarly equipped, could send e-mail messages and documents (research papers, experimental data) back and forth from Cambridge to Stanford. Stanford had a router, MIT had a router, and the NSF paid someone, such as MCI, to connect these two routers (and others) using a 56-kilobit-per-second pipeline—that is, an ordinary, if dedicated, telephone trunk line.

Because of the nature of a packet network, one user's router and data links may handle the packets of other users. If General Motors installs a router for its new plant in Moscow and pays for high-speed data links to other routers in Europe or the United States, General Motors will not be the only user of those facilities. Packets originated by Brooklyn AOL subscribers corresponding by e-mail with friends in Vladivostok will also transit these facilities. Thus the Internet is paid for by large users, especially ISPs, who want to connect their routers. Owners of very big routers on the backbone have to support the backbone links, and they charge others who wish to connect with their router or access point in order to support the expense of the backbone. But not all packets traverse the backbone, even those moving among LANs. Packets may go from one LAN to another within a given ISP's facility, or through direct connections between ISPs.

Long-distance telephone companies provide high-speed dedicated or leased lines to large Internet users and to ISPs. But more important, local telephone networks provide dial-up Internet access to ordinary consumers. This fact is of growing significance. The analog telephone system is constructed on the assumption that not everyone will try to use the telephone at the same time or for extended periods. A local switch that serves ten thousand customer lines ordinarily can service only about two thousand simultaneous calls. If more people than that try to call, they get no dial tone. Obviously, if a local exchange with ten thousand subscriber lines serves several hundred Websurfers who dial in and stay connected to the Web for hours at a time, there is likely to be a dial tone problem—for everyone.

There are two solutions: charge the AOL subscribers usage fees sufficient to curb their usage somewhat and to pay for new switching capacity, or invest in a larger switch. Usage fees have been ruled out by politicians anxious not to choke off growth, even unwarranted growth, of the Internet. At the same time, a larger switch makes telephone service more expensive, and someone will have to pay for it. Today telephone companies and ISPs are struggling over this point. The FCC will decide whether Internet users should pay for the costs they cause, or whether others should pay those costs. Because of the high hopes that the government has for the national information infrastructure, the official phrase for the Internet, it is unlikely that Internet users will be required to pay for all of the costs of the services they enjoy. (Kevin Werbach [1997] makes a number of convincing arguments showing that the ISP industry is not in fact being subsidized, or subsidized in any special way, and that there are also less costly technical solutions in which data transmissions bypass the local switch. The issue is not whether ISPs are being subsidized, however, but whether policymakers and the public believe that they are.)

As a consequence of the continued subsidization of Internet usage, whether from non-usage-sensitive access prices or from taxes on business and long-distance customers to promote universal service, the Internet will increasingly be used for services that are inappropriate. These are services that would not be provided on the Internet (or at all) if prices reflected costs. One example may be ordinary telephone calls. More than twenty different software applications and services now exist that permit Internet users to speak with one another "for free," bypassing the usual long-distance telephone system and its fees. Examples of such software are Intel Internet Phone, CU See Me, Televox, and Web Phone. At first limited to calls that had computers at each end, the software now requires no more than one computer-aided participant in what can be a multiparty conference call. Third-party providers such as AT&T Japan, Delta Three, and IDT offer telephone-to-telephone service. The network intelligence for such service resides in IP servers; in effect these companies are offering ISP services for specialized applications. It has been assumed that the major impetus for these services is the fact that international telephone calls are priced far above costs, and that domestic Internet providers escape the burden of certain cross-subsidies that are imposed on traditional telephone companies. There is undoubtedly some truth to this, particularly on the international level. However, it is also possible that IP network telephone service is actually cheaper to produce than standard telephone connections. This could be so for two reasons. First, the ISP industry is aggressive, innovative, and competitive in contrast to the traditional telephone industry. Second, "connectionless" IP networks make more efficient use of capacity than do networks with switched connections; at the same time, if there is a sacrifice of quality and reliability, that sacrifice is determined by the decision to rely on insufficient capacity and is not necessary or inevitable.

Left unchecked, and with the assumption that present quality problems are overcome, this and other forms of network interoperability may result in the disappearance of a large part of the demand for long-distance telephone service, as users take advantage of "free" Web calls. It still costs something to provide long-distance service on the Web, of course; indeed, many of the Web's facilities (and nearly all local connections) are provided by the transmission pipelines of the telephone companies. As a matter of fact, Internet engineers are hard at work installing systems to make the packet network more telephone-friendly. The protocol named RSVP (see Chapter 10), for example, reserves capacity in routers and transmission links to reduce latency, facilitating real-time exchanges. The same protocol may facilitate interactive video.

To the extent that the cost advantage of IP telephone service vendors arises from pricing distortions introduced by regulators, we can look for the FCC either to ban these services or (more likely) to impose taxes on such vendors that offset the effects of the distortion. This will mean, in effect, that certain uses (voice service) of the Internet will be taxed while others (data service) will not. From the point of view of the Internet, voice and data services are the same thing. Further corrective FCC taxes may thus be necessary. In short, it seems unlikely that the antiquated, inefficient and politicized telephone pricing system can exist side by side with an Internet pricing system that is market-based. Either telephone pricing will have to be rationalized or Internet pricing will have to be politicized. History teaches that the latter outcome is the more likely.

There is nothing more permanent than a temporary subsidy, nor anything less imbued with reason, especially in the communication business. For years, convoluted telephone industry pricing and other distortions have perpetuated subsidies that have no rational basis and for the most part never did. For example, the supposed telephone subsidies running from long-distance service to local service, from business service to residential subscribers, and from urban to rural callers for decades have been treated as sacred cows by regulators and legislators, to the point that policy changes designed to make the industry more efficient and competitive have been postponed or rejected whenever these subsidies are threatened. And yet there is little evidence that any of the subsidies accomplish even such simple objectives as benefiting needy people or other groups with any claim to benefit at their neighbors' expense. Raising business telephone rates, for instance, does not prevent that cost from being passed along to consumers and workers, with who knows what relative impact on the poor. Similarly, lowering residential telephone rates benefits rich and poor alike, and rural subsidies nowadays are as likely to benefit the owners of vacation homes as those enmeshed in rural poverty. As for long-distance rates, some studies have shown that poorer people make relatively more intense use of long-distance service than do those who are richer.

Commercial and political interests use these subsidies as a power base, and fight ferociously to keep them in place. Old subsidies, however unjustified or irrational, create for their beneficiaries an equity right in the status quo. Broadcast licensees and other spectrum users, residential telephone customers, and others who benefit from government-mandated subsidies will not easily give them up, and neither will the elected representatives to whom they make contributions.

There are signs the Internet may be headed down the cross-subsidy road. President Clinton's July 1, 1997, announcement that taxes on Internet commerce should be prohibited is an example. The hype and razzle dazzle of new technology promoted by commercial interests and inflated by visionary gurus like the pundit George Gilder catch the imaginations of politicians (even though Gilder himself is a free market advocate). This story is very much like the story of cable television in the 1970s, when that industry temporarily was the fair-haired child of government officials caught up in Ralph Lee Smith's 1972 vision of the "wired nation."

Predictably, the politicians want to make sure that the new technology does not merely benefit the rich and the information-elite—that is, those willing and able to pay for it. So programs are designed to bring the "information superhighway" to the urban and rural poor, to the bedridden, to elementary schools, to libraries, and so on, whether or not those targeted recipients have any use for it or value it more than other services that could be provided at the same cost. Because the government has no funds to pay for all this, the funding must come from a tax on other users of the Internet or the telephone.

The Telecommunications Act of 1996 contains provisions requiring the FCC to impose a tax on users of telephone and other communication services in order to fund subsidies that will ensure the availability of "universal service." It can only be a matter of time before the definition of universal service includes Internet access. Just as Marconi used the *Titanic* disaster to persuade Congress to mandate radios at sea and to promote radiotelegraphy at the expense of amateur transmissions, commercial interests that will profit from providing subsidized service will become powerful advocates of the continuation and expansion of universal service obligations.

Paying the Piper

Internet subsidies are hardly new. The whining of the early Internet users, for whom everything was "free" because someone else paid, has become tiresome. Commercial interests flooding the Internet face very real pricing questions on three levels: access, usage, and transactions.

Most ISPs and online services started out charging users a monthly access fee plus a per-minute or per-hour usage charge. This is a common mechanism for pricing telephone service, particularly cellular service. It is economically sensible if it approximates the structure of costs imposed by each user, and especially if it serves to ration use during those times when the Internet, or parts of it, are congested, with resulting delays for all users. It has two drawbacks. First, doing it correctly gets very complicated because usage prices should be zero except when the net is at capacity, and users ought to be able to pay different prices for different qualities or priorities of service, like first-class versus bulk-rate mail. Also, the "right" usage price is probably very small for services like e-mail and Web browsing, and if the right price is very small, then the cost of collecting it can inhibit efficient pricing. According to Jeffrey McKie-Mason and Hal Varian (in

McKnight and Bailey 1997, 49), the average cost of Internet backbone service per user per month is only about 50 cents. Congestion prices would be smaller.

The second drawback to usage-sensitive pricing is that people hate it. Possibly because few people are comfortable and familiar enough with Internet usage, or perhaps because so much online time is spent waiting for something to happen, usage-sensitive pricing makes the online experience much less pleasant for many users. For this reason, competition among ISPs has driven the industry to what may be, for the moment, an equilibrium pricing structure: fixed-rate access fees. Yet this has increased usage even more, creating further delays and capacity constraints.

Beyond access to and usage of bandwidth, there are interesting pricing issues related to Internet content. The "subscription mall" AOL model is one business plan, and the stand-alone proprietary site (for example, the *Wall Street Journal,* selling old articles for $2 each) is another. Both AOL and the *WSJ* charge a subscription fee for access and then a usage charge applicable to some but not all services—that is, a number of services are "free" to subscribers. The principal difference between the two models is the one alluded to above: malls work better than stand-alone stores in cases where the mall makes shopping more convenient from the point of view of time and travel, and in cases where the mall conveys a signal of credibility or quality to vendors and shoppers alike. Providing such services costs money, of course, and malls must compete with other distribution channels.

The high-priced services offered by the *WSJ* and some AOL and CompuServe premium "channels" are not representative of the Web. Most Web sites offer information (as opposed to services or goods). Much of this information is of low value, and much of it cannot be evaluated by the user until it has been consumed. If there are significant fees, whether for usage or access to such sites, traffic will cease. One business model for dealing with this problem involves secure payment systems and systems for paying small sums. Digital's "MilliCent" system and CyberCash are examples. Several systems for making more expensive commercial transactions on the Internet without credit cards exist, but all of them have elaborate encryption and other security features, making them unsuitable for transactions involving very small sums. MilliCent dispenses with much of the security and is optimized to deal with transactions under $5. The idea is that usage charges for site access would be so small—say, a penny or less per page per user—that wide-ranging Web browsing could continue, and yet a potentially substantial revenue stream for publishers would be generated. Users could very cheaply sample various sites and decide on the basis of actual experience which to patronize further.

The Demand Side

The real mystery of the Internet is not the pyrotechnic supply side of the market but the extent and nature of demand. No one will know the extent and nature of consumer or advertiser demand until it appears, or fails to appear, in the form of actual purchases of Internet services. What follows is speculation about what is in many ways the biggest unknown of all.

Consumers

To watch ordinary local TV channels, most people need do no more than purchase a color television set, available in small sizes for under $100, and extend the rabbit ears. In urban areas this will give them access to around 12 channels of television programs that various stations and networks pay more than $6 billion per year to acquire or produce.

Consumers craving more can add a VCR, for under $150, that lets them record TV programs to be viewed later, and also view rented movies and other video recordings. If they are still unsatisfied, 94 percent of all consumers can hook up to the cable television system running by their dwelling, and for around $24 per month increase to an average of 44 the number of television channels they can receive. Finally, insatiable viewers can sign up with a satellite service, for an initial fee of around $200 plus $50 to $100 per month, and get access to 200 television channels plus movies with closely scheduled start times (near video on demand or NVOD). For these four types of consumers, access costs range from less than $100 to several thousand dollars (for big-screen televisions with home theater features). Subscription and usage charges range from zero to $100 or more per month (for satellite TV with heavy movie viewing). There is hardly any consumer in the country that does not buy into the TV market at some level, and the same is rapidly becoming true throughout the world.

Access to the Internet with a dedicated phone line costs, at a minimum, $20 per month plus a computer. The cheapest computers designed to permit access to the Internet use TV sets instead of monitors and cost around $700, including modems. One such device, called Pippin, has a CD-ROM player, 5 megabytes of RAM, a 28.8-kilobit-per-second modem, a game paddle, and a keyboard. Hooked to a TV set and a telephone line, it connects with the manufacturer's ISP service, use of which is mandatory at $20 per month. It works and it is cheap, but it has several drawbacks: the installation process is complex and frustrating; the TV cannot be used to watch television, which has been a social or family activity, while Web surfing is, so far, a solitary activity; the telephone is tied up; and most Web pages are designed for monitors, making them hard to read on a TV screen. As a result, Pippin seems doomed.

Stepping up to more realistic Internet access—a basic personal computer with a hard drive, monitor, essential software, and adequate RAM—costs $1,200 to $1,500 or more. A second phone line costs $12 to $15 per month, and a modem costs $200. Everyone knows how to use a television set, and most people can install a new one in the family room unaided. Very few people over twenty-five are comfortable operating a personal computer, and installation can be a nightmare even for experts.

Having overcome the obstacles of installation and operation, the consumer finds that Internet access, whether through an ISP or an online service like AOL, eats up huge quantities of time, most of it spent waiting for information to be downloaded. Surfing is frustrating. Not only is speed a problem, but the home-page hype seldom matches the back-page reality. Not surprisingly, marketing experiments aside, the value of the information that is available for free is quite low. Material that might be more valuable typically requires surrender of one's credit card number, often in advance of any ability to assess the worth of what one is buying.

Different people have different tastes, of course. Many users who find surfing the Web frustrating find e-mail, chat rooms, and newsgroups to be an extremely satisfying Internet experience. But the Web browser is the frontier, and its promise and richness derive in great part from its graphical human interface—an interface that requires the transmission of much more information than e-mail and chats.

Consumers with unlimited budgets, or very strong demand for Web surfing, can get the telephone company to install a special digital ISDN line that sends and receives data up to four times as fast as the fastest modems in widespread use today. ISDN costs vary across the country, but a typical charge is $200 for installation, $25 per month, plus 0.5¢ per minute of use. In addition, one needs a "digital modem," which costs around $300. The installation process is even more of a nightmare than basic computer installation, chiefly because few telephone company installers have been trained in ISDN procedures, and none will help you actually connect the new line to your computer. Finally, having accomplished an ISDN connection, one discovers that many of the delays previously experienced had little to do with the speed of one's modem, but rather reflected congestion either at the server from which information was being requested, or on the Internet itself.

The Internet, in short, is far from cheap or user-friendly. And yet the success of the Internet is due to these same characteristics. In reality there is no paradox. The World Wide Web is cheap and user-friendly compared with the way it used to be with the Internet. But the World Wide Web is far from cheap and user-friendly, compared with using television. The trend is fairly clear: the World Wide Web is becoming, if not cheaper, a better value and increasingly user-friendly. But it is still interactive, not a passive medium like television, and there may in the end be no more demand for the Internet than there is for print media—itself a declining industry, because fewer and fewer people each year read books and newspapers.

For nonvideo Internet users, the Web is an exciting place despite its frustrations. Astounding numbers of people, firms, institutions, and governments have created Web sites and offer information. Rival search engines provide users with endless paths through the Web.

Advertisers

To an increasing extent, the Internet is supported by advertisers. Advertisers already provide substantial support to certain Web sites (for example, Yahoo, the most popular search site) and services (for example, AOL). Prospects for more ad revenue have been sufficient to float several initial public offerings of stock. Most newspaper Web sites, whether or not they charge user fees, rely on advertising revenues. By one estimate, Web-wide ad revenue in 1996 was around $300 million. Estimates for Web ad revenue in the year 2000 range from $1.7 to $5 billion. (Ad revenues of the broadcast television industry in 1996 were about $38 billion.)

Aside from some important technical details, Internet advertising is very much like other forms of advertising. It can be as specific and functional as a want ad or as image-oriented as a Coke ad. The advertising "banners" on Web pages that many users see may be nothing more than colorful brand names, or they may contain hyperlinks to other sites or order forms. It is the object of the Web-page publisher to expose as many users as possible to the ads, especially the kind of users likely to buy the advertised products. All this is the same as print and TV advertising.

The important technical details involve measurement. Print media must prove their circulations, and they are subject to check by the ABC (Audit Bureau of Circulation). TV broadcasters' audiences are measured by Nielsen and Arbitron. Advertisers make buying decisions and compare media based on each medium's cost per thousand (CPM) members of each advertiser's target audience. Thirty-second broadcast television ads typically cost from $6 to $14 per thousand viewers; four-color, full-page magazine ads cost around $8 to $20 per thousand circulation, and large daily newspapers generally earn $18 to $20 for ⅓-page black-and-white ads per thousand circulation. The Internet as yet has no standard way of measuring exposures or of auditing publishers' claims. Until such mechanisms are developed, Internet advertising will be suppressed. Estimates of 1997 Internet CPMs for banners range from $20 for Yahoo to $60 for ZDnet per page impression.

In spite of measurement problems, the Internet is attractive to many advertisers. For one thing, it has favorable demographics. Web users are young, well educated, and affluent—the kind of people who buy things. For another, the Internet ad provides a means for immediate gratification of impulses to buy, especially those that can be fulfilled by means of a hyperlink to another site. Even vendors of physical articles can immediately direct interested users to related Web pages and order forms, retail store addresses, and snail mail catalogs. The Internet is a hot medium.

Further down the road, if the Internet continues on its present trajectory, advertising on the Web and video advertising will become very similar in that both media will permit the delivery of audiences with demographics tailor-made for each advertised product. It is just a matter of the ratio of viewers to channels, and the increased content specialization that results when that ratio decreases. However, the availability of specialized content and fragmented audiences does not necessarily imply the end of mass audiences. They simply get more expensive for advertisers.

The Future of the Internet

One might think, from all the hype, and from seeing Internet addresses on the sides of every other plumbing truck and car wash, that most Americans are daily surfers of the net. The fact is that very few people surf the net. Doing so requires time, money, technical skill, and interests that the average consumer lacks. Users of the Internet, especially for services beyond e-mail, constitute a tiny fraction of the population. Although this fraction is surely growing, it remains highly unrepresentative of the population. Whether the Web becomes a mass medium does not turn on the preferences and characteristics of the elite group of current surfers.

The Internet must change if more households are going to use it. Indeed, the personal computer itself must change if more than about 40 percent of households are to adopt it. Today the valiant few who use the Internet must overcome barriers created by considerable expense, occult equipment setup procedures, obtuse interfaces, and often agonizingly slow responses. Clearly, they find significant value in the Internet to accept such abuse.

Michael Dertouzos, the head of MIT's computer laboratory, has trained many of today's leading software engineers. Yet he has become openly contemptuous of microcomputer interface design, claiming that it is the chief impediment to more widespread use of microcomputers. None of the improvements currently in the works seems sufficient to him; computer use, he claims, should be as easy as driving a car. That is not a very sanguine metaphor. It took many decades after its invention before most Americans felt comfortable operating an automobile. Except perhaps for automatic transmissions, power assists, and increased reliability, driving itself has not become easier. And to many Americans the automobile has been in important respects a social curse.

In its present form it seems extremely unlikely that the Internet will ever reach most households, much less become a mass medium like television. True, improvements in bandwidth, storage, and processor speed make the Internet more responsive to users. True, Moore's Law will reduce the equipment costs. But there is no reason to suppose that marked or sufficient improvements in interface design or content are in the wings. Meanwhile, television will continue to be delivered by means other than the Internet.

Conclusion

The Internet is the focus of a great deal of excitement, and for good reason. It is a novel communication medium and it provides a variety of services that were not previously available. The Internet is also the subject of hype, chiefly by commercial interests such as equipment manufacturers seeking to promote the Internet as a mass consumer medium.

Those who use the Internet heavily tend to be highly educated, affluent, and young. Much Internet use takes place outside the home, at offices and schools, where better connections, expert help, and faster computers are available. Such use is not, however, relevant to mass market acceptance.

Only a small minority of households make use of the Internet. This is explained by the expense of the equipment required and the difficulty in operating it. In addition, Internet service remains very slow and uncertain for most consumers because of the quality of local connections and other factors.

The Internet originated as a military project, and was later taken over by researchers sponsored by the National Science Foundation. It is now a mixture of commercial interests. Functionally, the Internet is an agglomeration of data networks using a common communication protocol. This protocol uses packets and routers rather than dedicated connections and switches. The Internet is optimized for transmitting e-mail and data files, not for real-time communications such as telephone calls or video broadcasts.

Until there are significant advances in technology, the Internet cannot provide mass market video services. It can, however, provide information and entertainment services that compete with the new digital television media. In order for the Internet to succeed as a mass medium in this sense, it will be necessary for the creative community to design program material and a user interface that is compatible with a mass market.

If the Internet does succeed in offering mass market services, both it and the capacious new digital video media will be moving away from the traditional consumer demand for passive entertainment service.

Actual or perceived government subsidies to Internet-related commercial interests may be the wedge that brings government into the Internet in the same way it was brought into radio broadcasting. Government mandates already exist requiring subsidies and cross-subsidies to certain groups to support Internet access. Some claim the ISPs are subsidized by other telephone subscribers under current pricing structures. All of this may lead to a perception by the public and policymakers that the Internet is a favored set of commercial interests. In that case, future regulation may arise, particularly if present expectations go unmet.

CHAPTER TWELVE

THE CB FAD: A CAUTIONARY TALE

In the 1960s and 1970s Citizens' Band radio took off. A service previously used by a few tens of thousands of people (chiefly truckers) suddenly attracted millions of ordinary consumers. By one estimate, there were 30 million CB users in 1977. As the number of users increased, the fixed bandwidth allotted by the FCC to this service became overcrowded. Too late, the FCC allocated additional spectrum. But the fad was already on the wane. Today CB is once again primarily a truckers' medium.

CB radio has the characteristics associated with network externalities. It is more valuable to each user, the more users there are. In the early 1970s, several unrelated events (a truckers' strike, gasoline rationing) gave CB a boost past the "critical mass" point needed to become a successful medium. But the boom went bust as the quality of service deteriorated due to congestion of the airwaves. The marketplace was unable to remedy this problem because of FCC regulations that prevented a timely increase in bandwidth.

Interestingly, CB fell from favor with the general public long before cell phones became widely available. That is, CB use did not fade merely because cell phones came along to offer the same service at higher quality.

There are parallels between the CB fad and the current Internet phenomenon. In each case the technology became an icon of pop culture and media interest. In each case the quality of service markedly declined as crowding increased the demands on the available bandwidth. In each case the crowding can be attributed to a flawed pricing scheme: users do not pay on the basis of how much or how often they communicate, but only on the basis of whether they communicate or not.

It seems entirely possible that most ordinary citizens eventually will give up waiting for hours while a modem slowly downloads a Web page that turns out to have, as its chief use, links to other Web pages and not the information originally sought. To sustain demand, some combination of the following must happen: the bandwidth must increase, so that the number of bits per second arriving at the home is significantly greater, or the quality of the information on the downloaded page must be greater, making it worth waiting for, or the price must be lower. The future of the Internet as a medium for consumer services and as a business opportunity depends on these factors. Absent changes in one or more of them, the Internet stands a real chance of being the CB fad of the nineties.

CB radio is not the only example of an ephemeral communication technology. The pace of change in communication technology has been so rapid that at any given time there are dozens of ideas, proposals, and projects, each backed by commercial (and sometimes government and academic) interest groups. Some of these succeed; most fail. In *Highway of Dreams,* A. Michael Noll (1997) points out that the term "broadband communications highway" originated in a 1971 report by the National Academy of Engineering. Most of its projections about the "wired nation" (at the time, a reference to cable television) have not yet come to pass. He cites many examples of failures (or in some cases very mixed successes). All of the following projects and proposals were highly touted at the time:

- Tele-education (1960s)
- Picturephone (1964–early 1970s)
- Teleconferencing (1970s)
- Tele-medicine (1970s)
- Warner Cable's Qube interactive cable system in Cincinnati (1977)
- Videotex (early 1980s)
- Integrated Services Digital Network or ISDN, a digital telephone technology that has now been overtaken by analog modems (mid-1980s–present)
- Bell Atlantic/Sammons's video dial tone (1992–1994)
- Bell Atlantic Dover Township's video dial tone (1996)
- Rochester Telephone's video-on-demand project (1995)

Even innovations that turn out well often take a long time to succeed. Consider the successful electronic innovations shown in Figure 12.1, and the number of years each required to reach even 25 percent penetration of U.S. households.

Figure 12.2 depicts "diffusion curves" for a variety of well-known household technologies. Even among technologies that ultimately become nearly universal, diffusion is sometimes slow. The telephone, whose diffusion was seriously slowed by the Great Depression, is an example. The telephone still has a lower penetration rate of U.S. households than the television set. Of course, many innovations, while not unsuccessful, never make it to universal acceptance. The VCR is an example. Similarly, on the basis of nothing more than the shape of their diffusion curve so far, there is reason to be skeptical that household personal computers in their present form will become universal.

In 1893, in the early days of telephony, it was widely predicted that the telephone would quickly revolutionize society, eliminate the need for office workers to travel to work, eliminate regional accents, and help eliminate strife and war. The telephone did have a major impact on society, but a rather different one than predicted, and much more slowly. If it were easy to predict how such inventions were going to work out, we would not need to pay entrepreneurs to take risks. At any given moment the air is full of possibilities, only a few of

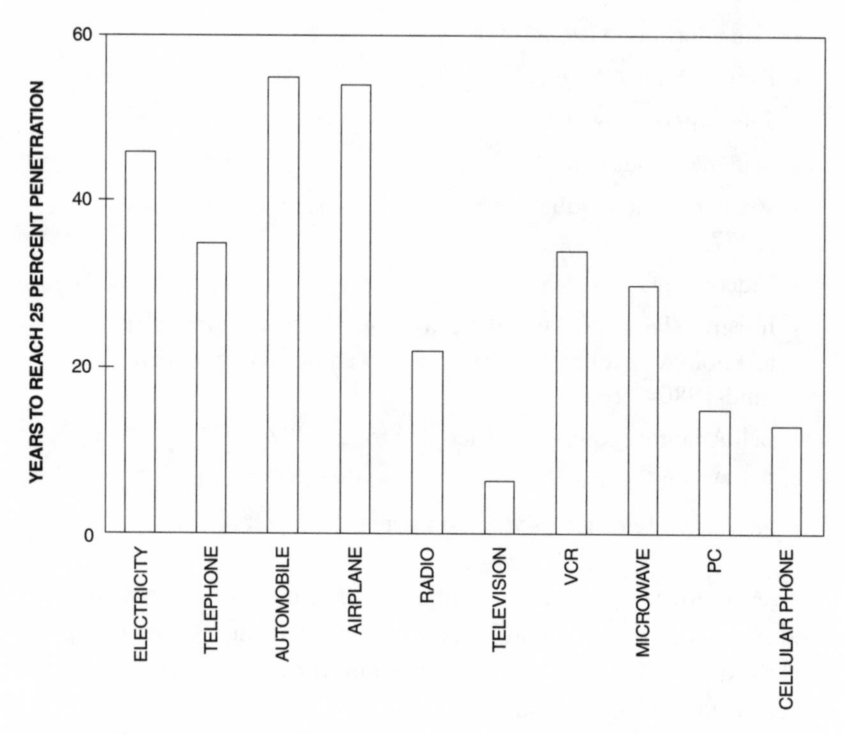

Figure 12.1 Diffusion rates of electric devices (*Wall Street Journal,* 1997)

which will become realities. Some possibilities are seen by politicians as so desirable that they are prepared to mandate their existence, with the understanding that someone else will bear the cost. Predicting which options will prevail is not a matter of deciding which technology is best—that is, fastest or clearest or cheapest. The best technology (viewed ex post) is hardly ever adopted, even by the market, much less by government agencies.

It is often very difficult in any event to say which is best, because the technologies differ in many dimensions. Technology is only the supply side of the equation. Demand matters equally. What services can each technology provide, and how much are consumers willing to pay for those services?

For forecasters, the problem is even worse than having to consider all the variations of supply and demand. Unfortunately, the technology that produces the greatest consumer surplus or value in excess

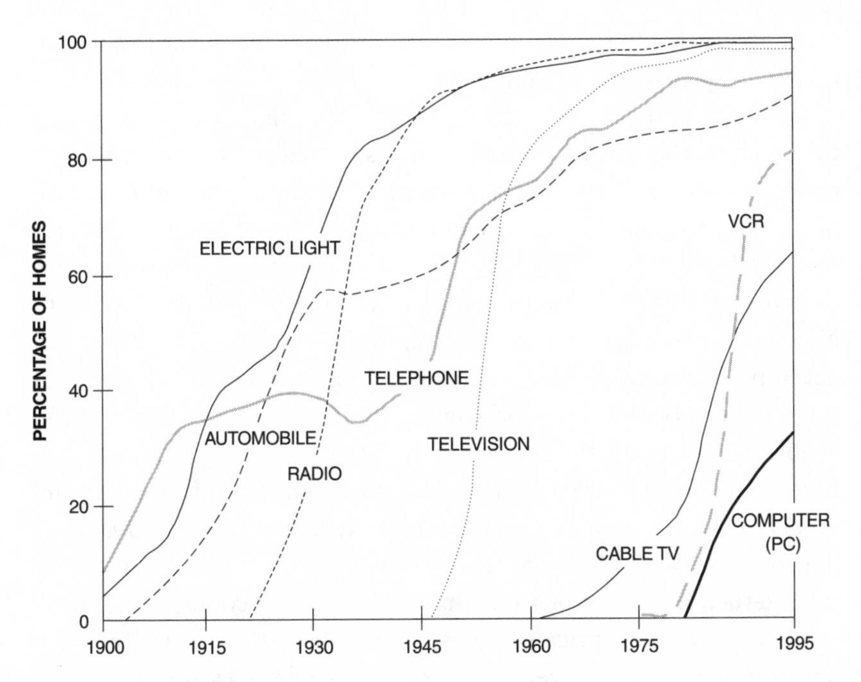

Figure 12.2 Diffusion of major consumer innovations: penetration of U.S. households, 1900–1995 (U.S. Department of Commerce, various years)

of its costs is not always the one that gets adopted. Technologies that are inferior in that sense may be adopted instead because they benefit from bandwagon effects or network externalities. One standard example is the (commercially unsuccessful) Apple Macintosh microcomputer operating system as compared with the (inferior) Microsoft MS-DOS (or Windows) system. Although it is widely agreed that the Macintosh is superior, and even though it had a "first-mover" advantage, it lost out because of Microsoft's superior marketing skills, and especially its good fortune in being retained by IBM to develop an operating system for a computer that most business users would want. A similar story is told about Betamax versus VHS videocassettes and players. The superior technology, bolstered with a first-mover advantage, lost out. Such market failure does not mean, however, that it would be better for the government to pick winners and losers. The government lacks any basis for making such choices successfully.

Moreover, it is often much more difficult to change government mistakes than private market failures.

Indeed, it is the importance of network externalities that accounts for much of the risk and excitement associated with each new computer technology. Consider the example of the Netscape Corporation's Web browser. Well aware of network effects, Netscape saw that success depended on two interlinked factors. It was important not only to have a large base of computer users browsing the Web with Netscape's products. It was also important to have a large base of developers producing Web sites that take advantage of the new features made possible by the browser. There was a chicken-and-egg problem, much like the one faced by early fax machine producers. Faced with the threat of aggressive competition from Microsoft and others, Netscape took a bold, risky, and expensive move: it *gave away* the client software. Millions of copies were offered on disk and over the Internet. It is quite possible that Netscape, if it could, would have charged a *negative* price—that is, paid people to use the browser. (Interestingly, in a different context, new cable television networks often pay cable operators an initial cash bounty to carry the channel; later the operator will pay monthly charges to the network, but only if the network is successful. This does not work with computer software because the seller cannot monitor the user to ensure that the product is actually used.) In any event the Netscape strategy seems to have been successful. Netscape is the dominant client browser software, and Netscape servers and associated software are dominant as well and at healthy prices. America Online followed a somewhat similar strategy when it distributed tens of millions of disks with its free software.

Conclusion

Although some new products and services are spectacularly and quickly successful (television, for example, or electricity), most take many years to reach mass market success. Some, like CB radios, are mere flashes in the pan despite vigorous publicity. There is very little evidence that either home use of personal computers in general or use of such computers for Internet access in particular is headed for immediate mass market success. Compared with other successful mass market products and services, the rate of diffusion of PCs is a disappointment. Although many people know about and may use the Internet, they do not do so from their homes in great numbers. Even if the Internet is headed for mass market success, experience suggests that it likely will be many years, even decades, before it is pervasive.

PART FOUR

FUTURE DIGITAL MEDIA

CHAPTER THIRTEEN

DIRECT-TO-HOME DIGITAL BROADCAST SATELLITES

"Starry night" may acquire a whole new meaning early in the next century as thousands of new communication satellites—nicknamed GEOs, MEOs, and LEOs—light up the sky. Some of these, especially the LEOs, seem very exotic. Most exotic of all are Alexander Haig's Sky Stations, hovering in the stratosphere over major cities.

Back in Orbit with GEO Digital Satellites

Chapter 8 described C-band satellites, the first communication satellites to broadcast TV signals in the United States. C-band satellites use analog modulation and relatively low retransmission power. Because C-band signals are so weak, their channel capacity is limited, and the required receiving equipment is relatively bulky and expensive. The next step in broadcast satellite technology has been digital satellites. Digital satellites use the higher frequencies at Ku-band. Like C-band satellites, Ku-band satellites are GEOs (geosynchronous communication satellites) and therefore have all the advantages and disadvantages associated with that orbit.

Ku-band satellites use frequencies of 12.2 to 12.7 gigahertz for uplink and 17.3 to 17.8 gigahertz for downlink; they also use much higher output power than do C-band satellites. As a result, the terrestrial antennas needed to receive their signals can be much smaller—as small as the 36-inch-diameter Primestar or the 18-inch DSS dishes in use today, for example. High-power Ku-band broadcast satellites use digital transmission. Both Japan and Europe had GEO Ku-band direct broadcast satellites for television before the United States. The first U.S. Ku-band digital broadcast satellite (1994) was a medium-powered "bird" and was owned by a consortium of cable companies named Primestar Partners.

As of 1995, according to the Department of Commerce, there were a total of twenty-five Ku-band broadcast television GEOs in orbit serving the United States. These satellites, together with fourteen older C-band and hybrid C/Ku satellites, have 604 transponders, corresponding to roughly 25.6 gigahertz of bandwidth. That is the same as 1,200 television channels (uncompressed, two per transponder) or 7,200 channels at a 6:1 compression ratio. Much of this channel capacity is used for private high-bandwidth connections, such as those linking TV news-gathering crews with their home studios.

Digital Ku-band DBS systems offer crystal-clear signals, even when converted to analog form for display on conventional TV sets; such systems also offer far more channels than are available on any cable system. The service is priced competitively with cable. The key questions regarding Ku-band direct broadcast satellites are these: How many competing systems will be economically viable? What effect will DBS competition have on the cable industry? How, if at all, will DBS operators solve the local content problem?

A Fitful Start

The direct (to home) broadcast satellite (DBS) business had a wobbly launch. At the beginning of the 1980s it was widely believed that DBS was an immediate and profitable opportunity. It seemed obvious that one satellite in the sky was a cheaper and more efficient way to reach U.S. television viewers than a system of hundreds of local stations, themselves linked with network headquarters by satellite connections. The argument for satellite distribution was even more compelling for regulatory reasons: there are no inherent limitations on the number of channels that a satellite can carry, whereas terrestrial stations are strictly limited. The FCC restricts the available frequencies, orbital locations, and maximum power of broadcast satellites, but it does not specify the technical standards for the broadcasts, which leaves DBS companies free to use digital transmission and compression to enhance their service. In contrast, the government restricts conventional TV stations in terms of frequency, antenna location and height, and power. It also specifies the technical standards that define the signal.

Around 1981, CBS, COMSAT, and Western Union, along with three other FCC licensees, announced DBS projects with much fanfare. By the mid-1980s all of the projects had failed or been put on hold, long before any satellites had been built or launched. Again in 1990 a project called SkyCable was announced, a joint venture by Hughes, Murdoch, and Cablevision. More than $1 billion was pledged by the partners. The deal fell apart within a year with no tangible result. USSB and Hughes in the same year signed an agreement that was to lead to the successful DSS (digital satellite system), and not long after the Primestar (originally K Prime Partners) joint venture was launched.

There are many reasons why these early projects did not succeed. First, technology until recently did not permit use of antennas sufficiently small to be practical in urban areas, and the FCC had not yet preempted local jurisdiction over the use of such antennas. Second, prior to digital compression technology, the channel capacity of the satellites was too limited to offer a striking contrast to terrestrial options, such as cable. Third, there was not enough programming available, especially in the early 1980s, to attract viewers; any successful satellite venture would have had to find or produce its own new programming. For these and other reasons it was not until Hughes's DSS system was launched in 1995 that a high-powered KU-band digital offering became commercially available. (Primestar was actually first up in 1994, but it is a special case, for reasons discussed below.) Fears, probably unwarranted, that cable networks might withhold programming (such as CNN or HBO) from DBS operators that compete with cable systems were quashed by the 1992 Cable Act. A provision of the law required those cable networks owned by cable systems to make programming available to all other media at nondiscriminatory rates.

A New Beginning

Today's digital broadcast satellites are superior to other methods of delivering television to the home in many respects. They offer more channels than are available over the air or from nearly any cable system. The quality of their pictures far exceeds that of off-air or cable broadcasts, even though their digital signals have to be converted into the old-fashioned standard NTSC format for conventional TV sets. For any given large number of subscribers, a satellite is cheaper per household, and much cheaper per household per channel, than cable delivery.

System costs for GEO direct broadcast satellites are high. Each satellite costs $150 to $200 million. Each launch (which can carry several satellites) costs between $50 and $60 million. In addition to the active satellite(s), spares are required either in orbit or on the ground. And development costs for a new system can run into the hundreds of millions. In spite of the enormous up-front investments required, satellite systems are attractive because of the dramatic economies of scale in serving large numbers of subscribers (see Table 13.1).

A system such as Hughes's DSS costs $750 million or more to launch, and each subscriber requires a set-top box that costs at least $200. Nevertheless, all it takes is 10 million subscribers to bring the capital cost per subscriber down to $275, and 10 million subscribers does not begin to exhaust the potential market of 100 million TV households in the United States. A large urban cable system has a construction cost of at least $600 per subscriber, maintenance expenses of about $60 per subscriber per year, and drop lines and set-top boxes that total another $75 or more per subscriber. Such a system might have 100,000 subscribers and therefore an overall capitalized cost per subscriber of $1,275. Not only is this number much higher than the satellite cost, but it cannot be reduced significantly through subscriber growth, because on average 70 percent of all U.S. house-

Table 13.1 Capitalized costs of satellite versus cable systems

	Satellite	Cable
Number of channels	200	60
Number of subscribers	10,000,000	100,000
Cost of fixed upfront investment	$750,000,000	$60,000,000
Subscriber equipment cost	$200	$75
Plant maintenance cost	0	$60/yr. or $600 capitalized
Programming	network	local and network
Capital cost per subscriber	$275	$1,275
Capital cost per subscriber per channel	$1.38	$21.75

Note: These are ballpark estimates.

holds already subscribe to cable service. The satellite cost advantage is even more striking on a per-channel basis.

Compared with any terrestrial system, wired or wireless, a satellite system has economic leverage because the incremental cost of serving a subscriber anywhere in the United States is close to zero, on-premises equipment excluded. Both wired and wireless broadcast systems must build new common facilities in order to extend their geographic reach or to add digital channel capacity. The satellite advantage thus becomes even greater if on-premises customer receiving equipment costs are minimized, perhaps by building much of it directly into digital TV sets.

The chief marketing disadvantage for DBS today is the difficulty in getting local programming. Television sets equipped with satellite receivers must also be connected to something else (rabbit ears, cable, or rooftop antenna) in order to receive local TV stations. DBS providers have tried to minimize this inconvenience by making available on the satellite a package of broadcast network signals (ABC, NBC, CBS, Fox, PBS). These signals are picked up from one or more affiliates of each network and rebroadcast under a statutory compulsory copyright license intended to provide for service to rural areas. It is widely believed that the service is also offered to nonrural customers, but if so it is unlikely that this unlawful activity will continue on a significant scale. In any event, this service does not provide local programming such as news and home-town sports. Many Ku-band DBS subscribers apparently continue to subscribe to basic cable service in order to receive local stations, adding around $16 per month to the cost of satellite service. In Table 13.1, adding $16 per month per subscriber ($192 per year, $1,920 capitalized) to the cost of satellite service changes the picture dramatically.

Viewers can receive local signals in other ways, including conventional antennas, that can be integrated with a DBS receiver. Some DBS systems are adding local TV reception capability to their antenna units, so that the integration can be seamless.

The key to complete success for DBS in a competitive struggle with cable (and other digital delivery systems, all of which are local) lies in the ability to offer local programming. In one sense this is just a matter of channel capacity and programming investment. After all, local-interest programming is no different than any other special-interest programming. The economics of the "Peoria Channel" are no different than the economics of the "Golf Channel." Special-interest programming of some quality can always be produced if there is enough subscriber or advertiser demand. As long as such demand is present, the only barrier to the provision of the programming is channel capacity. As long as channel capacity is present, the only barrier is program cost.

One way for a satellite carrier to address the problem of local programming is to produce its own local programming, at least in the largest cities, and provide a channel for each local-interest broadcast. Because those interested in such broadcasts are by definition local, it is possible to use spot-beam antennas on the satellite and therefore use the same frequencies several times. In other words, a "broad-beam" antenna would send national network signals to everyone, while several more narrowly focused spot-beam antennas would reach only specific regions, and each of the spot-beam antennas could use the same frequencies because the signals would not overlap. In this way, the channel capacity of the satellite could be multiplied several times without any increase in frequency use.

From the point of view of a satellite company, the preceding is not a very appetizing prospect. It takes a great deal of time and money to undertake production of local programming channels for dozens of cities, many of them in poor markets for advertisers. (One-third of all TV homes are in the top ten markets, two-thirds are in the top fifty, and almost 90 percent are in the top one hundred markets. The fiftieth market is Louisville, Kentucky, and the hundredth is Savannah, Georgia.) While on-air talent could be hired away from local TV stations without much difficulty, it would take some time to establish news-gathering staffs and equipment. Such channels would be at an initial disadvantage in bidding for sports rights and other popular local-interest programming, because local TV stations would have larger potential audiences. Although this strategy might work, and avoids various regulatory barriers, it would be expensive and risky.

Another approach is to pick up and rebroadcast existing local TV stations. This strategy offers several advantages. Local TV stations are already established and own valuable programming rights. A satellite audience is merely an incremental outlet for a broadcast station, and the price that such stations might demand probably would be low. But there are problems. Most important, local TV stations typically do not own the satellite distribution rights to their programming. A local broadcast station, when it buys a syndicated or network program, acquires only the local conventional television broadcast rights. Thus either the stations or the satellite company would have to negotiate for these rights with every program source. Even though stations might eventually begin to acquire such rights routinely, this problem poses up-front difficulties that may be comparable to those of simply starting local channels from scratch.

General Motors Strikes First

DSS (1994) is a joint venture of General Motors' Hughes unit (DirecTV) and Hubbard Broadcasting (USSB). As of March 1998, DSS had about 3.5 million subscribers. Both Hughes, a manufacturer of communication satellites, and Hubbard Broadcasting have had long-term interests in DBS. Hubbard applied for a satellite license as early as 1982. As noted earlier, Hughes was involved in the abortive SkyCable joint venture in 1990. Hughes was also a major supplier of the

C-band satellites used to transport video signals to TV stations and cable headends, and it pioneered the marketing strategy of selling individual C-band transponders rather than whole satellites.

GEO communication satellites are costly. Hughes, as a major manufacturer of satellites, seeks ways to promote their use. In this respect, Hughes's position is comparable to that of RCA in the 1920s. In order to sell radios, there had to be programming, and in order to support programming, there had to be radios. RCA sold radios and found ways to promote and to supply programming.

One way for Hughes to promote the sale of direct broadcast satellites was to offer a programming service that stimulated consumer demand. Even so, Hughes was unwilling to assume the risks of a DBS project without a partner. Hubbard took on some of that risk by contracting to cover the costs of about one-sixth of the DSS system capacity. Thompson, a French manufacturer of consumer electronics products and the current owner of David Sarnoff's RCA, agreed to pay the $150 million cost of developing the set-top converter boxes, in return for the exclusive right to sell them to the first million subscribers.

The Hughes DSS system cost $750 million, excluding development costs. It consists of three GEO satellites, all at 101 degrees west longitude. It has a total of 32 transponders, with power from solar cells at 240 watts per transponder. Each transponder can transmit 30 megabits per second. The frequency range of the satellites is 450 megahertz at Ku-band, 27 megahertz per channel, with two polarities. The system as a whole can transmit about 200 compressed video channels, depending on the material being compressed. The signals are received by an 18-inch dish connected to a TV set via a converter box that costs several hundred dollars. The receiving equipment must be mounted on the south side of a building or other structure with a clear view toward 101 degrees west longitude. For the moment, this leaves out homes surrounded by hills or dense foliage and apartments not on or lacking access to the south sides of buildings. (Whole apartment buildings can be served using a single rooftop antenna and interior coaxial wiring.)

The DSS equipment handles both DirecTV (Hughes) and USSB. USSB uses its five transponders to supply premium movie services, such as HBO. The bulk of the regular cable networks (CNN, Lifetime, Discovery, and so on) are on DirecTV, which also offers pay-per-view movies, premium sports channels, and a multichannel Muzak-type service. The movies are offered on a "near video on demand" (NVOD) basis. Each movie is repeated on several channels with staggered starting times. The most popular movies start every half-hour. USSB and DirecTV sensibly do not duplicate programming, although the movies shown on USSB premium channels, such as HBO and Showtime, often duplicate the offerings on DirecTV's pay-per-view movie channels. About half of DSS subscribers take both USSB and DirecTV service. The balance take just DirecTV.

More than half of DSS subscribers are (or were) cable subscribers as well. DSS offers far more channels than most cable systems, with far higher reception quality. The monthly subscription rates for comparable packages is roughly the same on cable and DSS. Of course, consumers must invest $200 or more in DBS receiving equipment that would not be needed with cable or that would be provided by the cable operator on a rental basis. Many high-end cable subscribers find they have more choices and better reception on DSS, and they are willing to pay more for it. To retain local broadcast signals, many continue to subscribe to basic cable service.

Converter or set-top boxes (and associated antennas) serve two functions. First, they receive the digital signals and translate them into analog signals that can be fed to a conventional TV set. Second, they contain circuitry that prevents use by those who have not paid for the service, generally by showing scrambled signals to nonpayers.

The problem of financing consumer receiving equipment is an important one. If a converter box costs $200 and if all 100 million TV households need one, the aggregate cost is a staggering $20 billion plus installation costs, much more than the cost of the satellites themselves. Further, every delivery system today has a different converter box and antenna, so consumers generally cannot switch from one satellite service to another or between cable and satellite without making the investment anew. Several DBS competitors now offer the equipment—converter and antenna, but not installation—for around $200, tied to a year's subscription to the programming costing about $300. One service, Primestar, includes the converter box in the price of program service, minimizing the consumer's up-front cost. Cable operators also finance the box for their subscribers. It seems likely that the trend will be toward increased supplier financing of converters, just as cellular telephone services tend to finance the purchase of cell phones by offering deeply discounted or even free handsets in return for a minimum commitment to purchase service. This strategy, akin to the free software distributions by Internet service providers, helps suppliers ride the network effects wave and move rapidly down the economies of scale curve. Of course, it also magnifies suppliers' own financing problems; finding a way to finance these boxes is a big hurdle for potential entrants such as Rupert Murdoch. David Sarnoff would have recognized the problem.

Eventually, as the new digital television sets become available, common circuitry may accommodate any of the services without the need for a converter box. The FCC will be under continuing pressure to impose such a requirement on manufacturers in order to avoid the need for three or four separate boxes on top of each TV set. Imposing a mandatory standard may reduce direct and switching costs, but only at the risk of adopting and locking in a mistaken standard.

Other Stars in the Sky

Primestar is a 120-channel joint venture of Comcast, TCI, Continental, Cox, and Time Warner, all large cable MSOs (multiple system operators). As of March 1998, Primestar had 2.0 million subscribers.

Primestar, not DSS, was actually the first U.S. digital Ku-band broadcast satellite. Because Primestar was only "medium" powered, receiving dishes had to be rather large—around 36 inches in diameter—though still less than half the diameter and one-fourth the area of C-band dishes. Nevertheless, a 36-inch dish was enough to limit Primestar's appeal in urban areas. Further, its cable MSO owners saw Primestar chiefly as a means to bring video services to rural and other areas where it was uneconomical for them to wire. Accordingly, cable systems were often responsible for marketing the service in each locality. Cable systems did not promote the satellite service to existing cable subscribers, at least as long as Primestar was the only satellite service. The Department of Justice, the Federal Trade Commission, and various state attorneys general have repeatedly scrutinized Primestar's structure and ownership on this account.

As DSS proved, cable subscribers who are heavy television viewers are the most likely early customers for DBS. More than 90 percent of all households have the option of subscribing to cable, but only about 70 percent do so. The rest are relatively poor candidates for a new delivery system that generally offers more video options at a higher price than cable. Many cable subscribers, in contrast, are willing to pay extra for even more choices and even better reception, just as they did when they first subscribed to cable.

In 1997 Primestar acquired a high-powered Ku-band bird, though one with limited channel capacity. This satellite has 80 channels. A new 30-inch dish permits reception of Primestar's new and old satellites, so it now has a total of about 200 channels. When Murdoch's adventure with the Sky project ended, as described below, Primestar agreed to acquire Murdoch's satellite assets. This deal may or may not win approval from antitrust authorities.

EchoStar, which operates the DISH Network satellite system, was founded in 1980, but had to wait sixteen years to launch its first satellite. The two-satellite DISH system at 119 degrees west longitude was launched in 1996 and by March 1998 had about 1.1 million subscribers. It offers 200 channels and programming comparable to that of DirecTV/USSB and Primestar. Subscribers use an 18-inch dish. In 1996 EchoStar bid $52 million for an additional satellite orbital slot at 148 degrees west longitude, from which about three-quarters of the U.S. population can be served, omitting the eastern seaboard. In addition, EchoStar owns a slot at 61.5 degrees west longitude and another at 175 degrees. The two additional slots are slated for rebroadcast of local TV signals, if EchoStar can overcome its difficulties.

EchoStar suffered a blow in 1996–1997 when it agreed to merge into Rupert Murdoch's ASkyB system to form Sky. The deal came apart within a few months. Charlie Ergen, EchoStar's chief executive officer, was left to pick up the pieces of a DBS system that had always struggled to find adequate financing. One asset EchoStar has is a multibillion-dollar lawsuit against Murdoch.

Rupert Murdoch Fumbles the Bird

Rupert Murdoch, whose media empire includes not only newspapers, magazines, and the Fox TV network but direct broadcast satellites serving Europe, Latin America, and parts of Asia, has been poised to enter the U.S. direct broadcast market. Faced with the fact that three systems (DSS, EchoStar, and Primestar) were already operating in this market, Murdoch tried to assemble a package that could compete successfully not only with terrestrial distribution systems but also with existing satellite companies. The key to his strategy was local content.

Murdoch started 1997 with ASkyB, a direct broadcast satellite under construction. (ASkyB stood for American sky broadcasting, a play on Murdoch's successful BSkyB, British sky broadcasting.) ASkyB was owned by Murdoch in a joint venture with MCI Communications (then slated to be acquired by British Telecom). ASkyB was to utilize an orbital slot at 110 degrees west longitude which the FCC decided to auction off and for which MCI paid nearly $700 million. According to the original plan, Murdoch would operate the video entertainment side of the venture, while MCI would use some of the planned capacity to offer business-oriented services. Murdoch's opponents tried to use MCI's proposed acquisition by British Telecom, together with U.S. laws and regulations disfavoring foreign ownership of U.S. telecommunications assets, to delay the project.

The key to the venture was local signals. There are about 1,600 local TV stations in the United States, and it is not economically feasible to rebroadcast all of them by satellite to their local viewing areas. (To do so is certainly possible. Given the available bandwidth, some combination of advanced compression, increased satellite broadcast power, and increased receiving equipment sensitivity could do the job. It simply would cost too much at present to be commercially viable.) What Murdoch needed was an additional supply of bandwidth and orbital slots.

EchoStar's rich supply of orbital slots and potential capacity made it an obvious target for Murdoch. In February 1997, EchoStar agreed to merge with ASkyB to form "Sky." By April 1997, the deal was off and the parties were suing each other. Murdoch had fumbled badly in his lobbying efforts to get favorable legislation through Congress. Nevertheless, the ambitious dimensions of Sky are worth recounting. They shed light on business strategy in the competition between satellite and terrestrial digital broadcast delivery systems.

Sky would have had eight high-powered Ku-band satellites in two of the only three available orbital slots that have full coverage of the continental United States. Several of the satellites would have had "spot-beam antennas" designed to provide retransmission of major local broadcast stations within relatively small areas around each of the largest television markets. Eventually, Sky hoped to deliver local signals to 75 percent of all television households. In all, Sky would have had 50 transponders using an (average) 10:1 compression ratio, and thus 500 channels, not counting channels created by frequency reuse.

The Sky business plan emphasized local signals, the distinguishing selling point of cable television in its early struggle with DSS. Like cable, Sky planned to finance most or all of the cost of customer equipment, such as set-top boxes and receiving dishes, so that up-front cash layouts by consumers would be minimized. According to press reports, Sky expected a minimum of 8 million subscribers by its fifth year of operation. This would have required an investment of around $3 billion, or more if all customer equipment were financed.

The Sky project failed for many reasons, dissension among the partners among them. But the key element probably was Murdoch's failure to obtain legislation giving him a free compulsory license to retransmit local TV stations within the stations' own coverage areas. Cable operators now have such a license for local broadcast signals, but they are required to carry all the local channels, under the so-called must-carry rule upheld by the Supreme Court in 1997. Congress apparently would grant Murdoch this dispensation only if he agreed to carry all 1,600 or so local stations, something he could not afford. The cable and broadcast industries fought Murdoch ferociously. They would allow Sky only if it went forward on the same terms and with the same obligations as cable systems. A political coalition of broadcasters and cable operators makes a formidable opponent even for Rupert Murdoch.

Murdoch faced two other complications in carrying out his plan. One was his simultaneous effort to launch two Fox cable channels he owned. The success of this expensive effort depended on the willingness of cable MSOs to carry the new Fox channels. Sky was a spear aimed at the heart of cable, and MSOs did not rush to subscribe to the new cable services. A second complication was how to integrate his Fox television network and Fox-owned stations. The Fox-affiliated stations, most of them independently owned, were naturally concerned with the ability of a Fox DBS system to bypass them, beaming programs and advertising directly to viewers. Murdoch's plan to have a large number of spot-beam antennas provide local station retransmission to three-quarters of all viewers would potentially leave one hundred or more Fox affiliates facing competition from their own network. (The affiliates facing competition would be those in markets too small to warrant use of the capacity-constrained spot-beam transponders. The other broadcast network affiliates will have a similar problem.) Murdoch's proposal to deal with this issue was to have decoders (set-top boxes) block satellite-relayed broadcast signals that did not originate in each subscriber's local market.

The details of Murdoch's debacle are not important. Neither he nor the other players are finished with this game, and nothing resembling equilibrium has been reached in the direct satellite broadcast business. The point is to illustrate the tactics and processes by which commercial success must be pursued, and the enormous stakes. In the end, it may be that three DBS systems rather than four or five will survive, but which ones or who will own them is very much in doubt.

Solving the Local Content Problem

The local content problem probably will have to be solved in some other way than the Sky approach. One possibility is for Murdoch's News Corporation or other satellite companies to make deals with local stations under which the stations would produce special satellite editions of their regular broadcasts; substitutions would be made for programs where satellite rights were lacking. Station revenue would come from increased advertising and possibly from payments by Murdoch or other satellite companies, just as cable networks and other special-interest program sources are paid. As noted earlier, there really is no difference between a local-interest channel and any other special-interest programming, from an economic point of view. Or if there is, it lies in the differing regulatory treatment of the two. The Cooking Channel is completely unregulated, while TV station WCBS is as heavily regulated as many public utilities.

There is a range of other possible solutions to the local content problem. One is that demand for local content may simply wither away as more and better national special-interest programs become available. In Europe there is much less local programming; programming tends to be regional or national. In the United States, very popular local content might migrate to national channels. Right now, for example, DirecTV offers special premium packages that cover all or a large percentage of professional football and basketball games, using several channels to do so.

Another solution is that existing local TV stations will be delivered over the air in a digital form that makes integration with a satellite receiver trivial. It is only the awkwardness and contrasting quality of reception that make integration of a satellite receiver and a rooftop TV antenna challenging today. As digital broadcasting becomes more widespread in America over the next ten years, this problem may disappear. This possibility has momentous potential for increasing competition in digital television delivery.

Lowering One's Sights: Low Earth Orbit (LEO) Satellites

Low Earth orbit (LEO) satellites, unlike GEOs, don't stay put 22,300 miles above a spot on the equator; instead they zip along only a few hundred miles up. In 1997 the ashes of the hippie guru Timothy Leary were launched into such an orbit from a high-flying airplane. Because LEOs are moving relative to the Earth, a great many of them are required in order to ensure that at least one will always be in sight of a given receiving device. The lower the orbit, the more birds are required, because the footprint of each bird—the area of the Earth within its horizon—decreases as it gets closer to the Earth. (If a satellite actually fell toward Earth, its footprint would get continuously smaller until it approached zero at the point of impact.)

While low altitudes require LEO systems to have more satellites, they also reduce the power requirements both for the satellites and for the Earth-bound receiving devices. LEOs require less powerful launch vehicles than do GEOs, and some (such as the one with Leary's ashes) have been launched from high-flying aircraft. As with spot beams on GEOs, LEOs can reuse frequencies and thus transmit more data on a given allotted bandwidth. Finally, LEOs reduce the time delay or latency in communication, making them more suitable than GEOs for interactive services. LEOs are at a disadvantage, however, in requiring a large number of satellites and hence launches. Indeed, presently proposed LEO systems would require substantially more launch capacity than the world possesses. Further, much of their capacity is wasted because so much of the Earth is water and wasteland. Furthermore, the need to provide for hand-offs of ongoing connections from one satellite to another complicates the engineering problems.

The first communication LEOs were used by the former Soviet Union as part of INTERSPUTNIK, its global communication system, and were designed to accommodate high latitudes where GEOs above the equator are less effective. (The antenna on an Earth station at high latitudes must face very close to its horizon to see a GEO satellite; this leads to increased interference and other problems. Russia has a significant amount of territory at high latitudes.) Modern commercial LEO systems have been motivated chiefly by the needs of mobile telephone users. The idea is to bring satellite-based cellular telephone and paging service to the entire globe.

There are big LEOs and little LEOs, based on the amount of spectrum each type has been allotted. Little LEOs seek to offer global services such as two-way paging. Several of these systems are now in service: Orbcom, Starsys, and VITA. Others are being planned. Qualcomm has offered a highly successful low-speed little LEO data service, OmniTRACS, to truckers since the late 1980s. Little LEOs have very low capacity and lack video or Internet applications. Big LEOs are still in the design stage or under construction. They will offer worldwide mobile telephone service not unlike present-day cell phones. The three leading big LEO systems are Iridium (Motorola and others), GlobalStar (Loral, Qualcomm, and others), and Odyssey (TRW and TeleGlobe Canada). All three are focused on mobile telephone service and seem to be on track to launch service before the millennium. None is likely to be able to supply significant capacity for video or Internet use.

Iridium (named after the seventy-seventh element in the periodic table) was the earliest and perhaps the most ambitious big LEO project. Motorola originally proposed launching seventy-seven LEO satellites to provide a global mobile telephone system. The $5 billion Iridium project (now scaled back to a constellation of sixty-six satellites) was launched in 1997–98, and was to have begun service late in 1998. The satellites were launched from facilities in Russia, China, and the United States. Financial arrangements with the foreign communications companies that will market the service in their respective countries have reduced the business risks faced by the venture. Iridium satellites orbit at an altitude of 488 miles.

Given current compression technology, neither Iridium nor the other first generation big LEOs have the capacity to provide video services, and they would have only limited capacity for data transmission (about the same as ordinary telephone lines using obsolescent 14,400-bit-per-second modems). Thus neither big LEOs nor little LEOs have any immediate implications for television delivery or broadband Internet services. LEOs are of interest chiefly because future generations of this technology may be important components of a broadband low-latency network. And, of course, lack of bandwidth is not an absolute barrier to video and similar services because processing power and storage can be substituted for bandwidth.

Teledesic, SkyBridge, Celestri: Internet in the Sky

In a class by itself when first proposed was the $9 billion "broadband LEO" (also called a "mega-LEO"), conceived by Craig McCaw and his partner, Bill Gates of Microsoft. Functionally, if Iridium is a cellular phone system in the sky, Teledesic is the Internet in the sky. Unlike the other LEOs, Teledesic and its imitators are designed for fixed rather than mobile users, and for data rather than voice transmission. Teledesic is direct-to-home, and two-way. Where Iridium uses 66 LEO satellites, Teledesic proposes to use 288—plus spares. Indeed, Teledesic's original plan involved more than 800 satellites. (See Figure 13.1.) The plan is to launch these satellites—approximately one per day—into orbits at altitudes below 500 miles during a twelve-month period ending in 2002, a feat that apparently will by itself require half of the world's rocket launch capacity. Boeing, the satellite prime contractor, has taken an equity interest in Teledesic. Teledesic is able to deliver greater capacity than Iridium because of its larger number of satellites, each serving a smaller geographic area; in other words, the bandwidth is reused more intensively, as in a cellular telephone system with many small cells.

Teledesic is designed to operate in the Ka-band (28 to 29 gigahertz), which it shares with LMDS (local multichannel distribution service, discussed in Chapter 14); the FCC has recently had to settle disputes

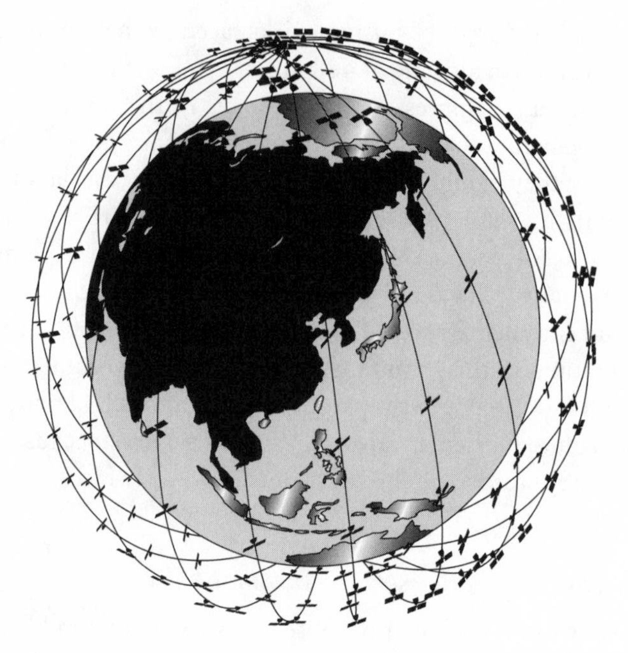

Figure 13.1 Teledesic's satellite constellation (Teledesic LLC, 1998; used by permission)

between terrestrial and satellite users of this band concerning interference standards. The Teledesic satellites will speak to one another as well as to Earth, introducing Internet-like redundancy and reliability. Because their low orbits reduce transmission delays and latency, these satellites are suitable for interactive services such as real-time video and audio. The earliest LEOs required steerable antennas at the receiving stations on Earth to track the satellite as it moved across the sky and to switch to the next one as it came up over the horizon. The Teledesic system has (virtually) steerable antennas both on the ground and on the satellites. The antennas can be focused electronically rather than mechanically, which will make them less expensive and more reliable.

According to its own Web site, Teledesic is designed to serve "millions of simultaneous users." "Most users will have two-way connections that provide up to 64 megabits per second on the downlink and up to 2 megabits per second on the uplink" (www.teledesic.com, April 9, 1998). In addition, it can serve several thousand high-volume users (such as interconnecting telephone companies or Internet gateways) at a 1.24 gigabit-per-second rate. Teledesic will be a packet network with asynchronous transfer mode (ATM) switching, the next step in local telephone digital networks. A characteristic of the technology is that downstream (Internet-to-user) data flows are much greater than upstream (user-to-Internet) flows.

The Teledesic plan is breathtaking in its scope and ambition. When first announced it seemed almost quixotic. But the nearly immediate response by two powerful competing groups (SkyBridge and Celestri) lent credibility to the concept if not to the details of the Teledesic proposal. Mega-LEOs are the only satellite systems now planned that might have an effect on the issue of convergence of computers and video media. According to McCaw, the Teledesic system can be expanded "gracefully" in the future. But it is clear that at planned levels of capacity, even mega-LEOs such as Teledesic cannot serve as a mass market broadband connection to the home. Despite higher maximum rates, Teledesic's planned per-user data rate is 2 megabits per second, less than what is required for standard-quality real-time video with present-day compression technology. (Recall that DirecTV uses 3 megabits per second when broadcasting its pay-per-view movies.) Moreover, Teledesic cannot supply even 2 megabits per second to all 20 million users simultaneously. Finally, while 20 million users is a big number, there are 95 million television households in the United States alone, approximately 30 million of which have computers. In the broadcasting world, a maximum user base of 20 million is not enough to compete at the highest level. At the same time, if three or more competing mega-LEOs are to be launched, aggregate capacity begins to look more impressive, and prices may be more affordable.

Beyond Teledesic lies a range of largely hypothetical satellite systems using Ka-band technology. Hughes, Motorola, AT&T, GE, Loral, and nine other companies have applied for a limited number of orbital slots for very high capacity GEO communication satellites that would in principle support two-way telephone and Internet connections from the home. Most would have the capacity to provide interactive video service, although overall capacity (number of simultaneous users) for such service would be limited. The Ka-band satellites, including Teledesic, contain a special feature: on-board digital signal processors that act as electronic switches. A given user can be connected with any other user, including one in a different spot beam of the satellite and even one served by a different satellite. Such satellites are no longer mere transmission trunks; they are switched telephone systems.

Me Too: Celestri and SkyBridge

Teledesic did not remain unique for long. Early in 1997 the French firm Alcatel (with partners Loral, Toshiba, Mitsubishi, and Sharp, among others) announced plans to build still another mega-LEO called SkyBridge. As proposed, SkyBridge would compete directly with Teledesic and terrestrial media for Internet business. SkyBridge is slated to cost $3.5 billion and to be operational with sixty-four satellites in 2001, before Teledesic. SkyBridge operates in the Ku-band and is less complicated than the other mega-LEOs, lacking, for example, intersatellite links. On the other hand, SkyBridge has proposed controversially to use frequencies already licensed exclusively to others, and to avoid resulting interference by automatically turning off its own signal (or switching frequencies) whenever interference threatens.

In June 1997 Motorola responded to Teledesic, and especially to Teledesic's choice of Boeing as a construction partner, with its $12.9 billion Celestri proposal. Celestri involved a complicated combination of several GEO and sixty-three LEO satellites, operating at an altitude of 900 miles, and in the same frequency bands that Teledesic already shares with wireless cable systems. The satellites were to have on-board ATM switching capacity and to communicate among themselves using optical wavelengths. To obtain its license, Motorola sought to demonstrate that its system would not interfere with Teledesic or with terrestrial LMDS wireless cable systems using the Ka-band. The earlier entrants have an opportunity to delay Motorola by claiming interference. This provides Teledesic with a competitive advantage and with little incentive to share spectrum resources.

Although Celestri was slated to serve the same business and data communication markets as Teledesic, it was also aimed at home video entertainment. The proposed system contemplated uplinks ranging from 64 to 2,048 kilobits per second from residences, with "bursty" downlinks to subscribers at up to 10,000 kilobits per second for data and 2,000 for video. The peak data transfer capacity of each satellite was to be 17.5 gigabits per second, which translates into a maximum of 400,000 simultaneous 64-kilobit-per-second channels in the continental United States, and 1,800,000 such channels worldwide. A 64-kilobit channel is not very big, corresponding roughly to today's fastest consumer modems, or the theoretical data transfer capacity of an ordinary analog telephone voice connection. Thus Celestri would have been able to serve approximately 1 percent of the current demand for telephone connections in the United States. It would have been able to serve a much smaller share of the demand for television service. If systems such as Celestri turn out to be the most cost-effective way to provide interactive video and residential data service, a great number of them will be needed. But Celestri itself will not be one of them. Its owner, Motorola, bought a 26 percent stake in Teledesic in May 1998, canceled Celestri, and took over the job of being Teledesic's prime contractor.

I Am in Charge: General Haig Sends Up a Trial Balloon

Just when transmission technology seemed to be focusing on its next major change, from fiber-optic cables to satellite transmission, a new idea has emerged: balloons (dirigibles, really) in the sky. A startup company headed by retired general Alexander Haig and members of his family announced plans in 1997 for Sky Station, a network of 17-ton, 450-foot-long, lighter-than-air craft that would hover above 250 or more major world population centers at very low altitudes even by LEO satellite standards—about 18 miles (see Figure 13.2). Sky Station would provide high-bandwidth, last-mile communication services. This system (based on a defunct government scheme to fill holes in the ozone layer) may be simpler and less costly than either LEOs or GEOs because it does not require satellite-to-satellite connection and switching (although some of that could be done), and it does not require the construction of elaborate facilities whose capacity and power are largely "wasted" serving the oceans and uninhabited tundra or deserts. In addition, Sky Station does not require expensive launch vehicles. On the other hand, use of an experimental slice of the spectrum—47 gigahertz—and an experimental ion engine running off

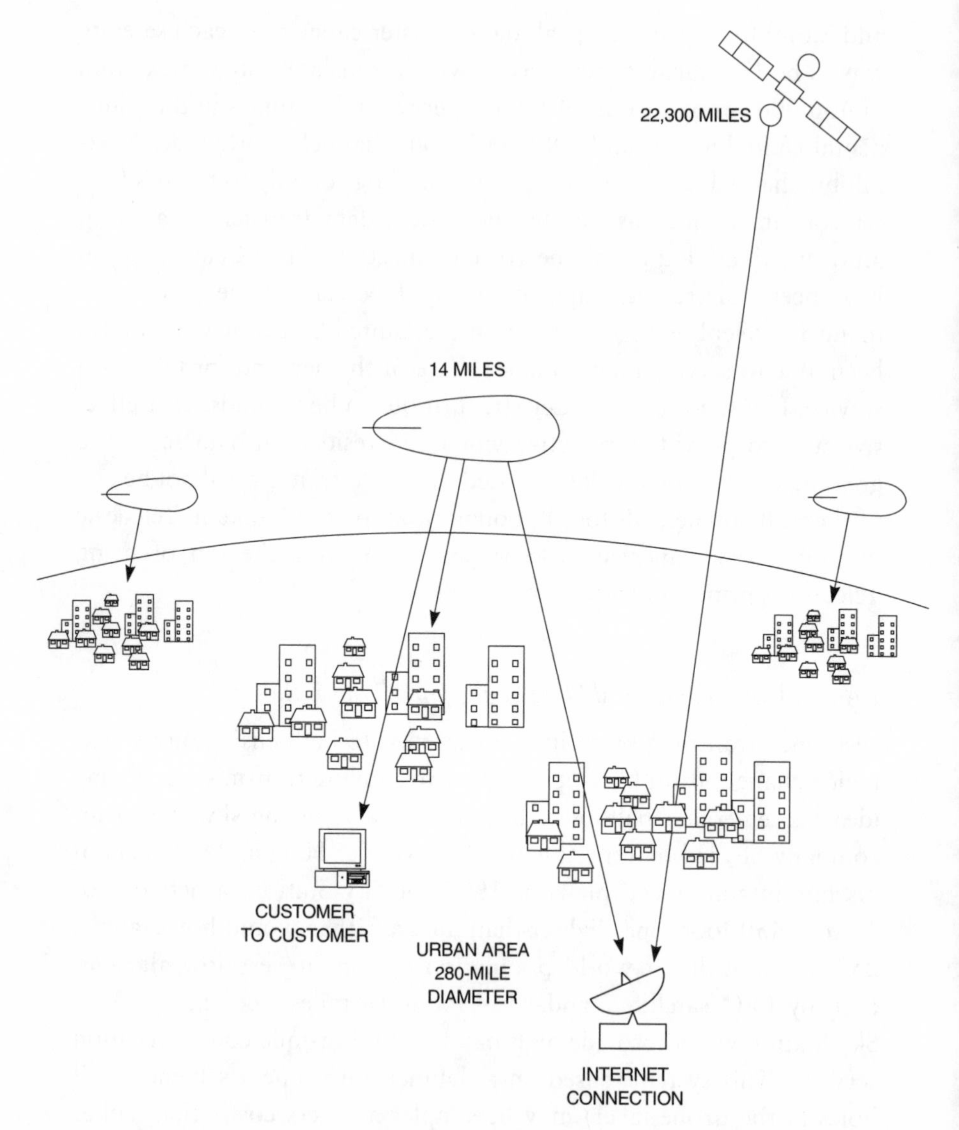

Figure 13.2 Sky Station

solar panels to power the dirigible may present challenges. (Another aspirant has proposed a similar scheme, using piloted aircraft kept circling each city in round-the-clock shifts.)

Sky Station does do one thing exactly right: unlike any of the satellite systems (except perhaps SkyBridge), with proposed 64-kilobit-per-second links it focuses its sights on the point where the problem lies—the "last mile" to the home. (The "last mile" is the local loop that connects household telephones to telephone company switches, and which is everywhere a monopoly of the local phone companies.) At present, of course, this relatively modest capacity rules out the use of the Sky Station for television transmission.

Satellites and the Future of Television

Do satellites have a future in television, either as backbone trunks or as transmission paths to consumers? Digital GEO Ku-band and Ka-band satellites will remain an extremely efficient and flexible means of transmitting to the home. They also have an inherent cost advantage at a scale that no terrestrial system, wired or wireless, can match. Prices are already competitive with cable system prices and will continue to fall as penetration and competition increase. Yet in spite of their advantages, GEO satellites face serious difficulties. Most formidable is their inability to offer cheap local content (or, more generally, their inability to offer enough channels so that parochial interests can have their own "network"). This problem may be solved in one of the ways discussed earlier, and if it is, satellites could dominate the business of video-to-home broadcast transmission for the foreseeable future. Satellites, however, are subject to regulation because of their use of government-controlled orbital slots and frequencies. Given the ease with which the government can be influenced, especially by the broadcasting industry, this does not bode well for the satellite companies. The regulatory process gives competitors, both cable and conventional broadcast, leverage to slow or stop satellite success. In preventing Murdoch from achieving his goal of a free compulsory license for local TV signals, they have already accomplished much.

Digital GEOs seem unlikely candidates to supply two-way interactive video services or Internet content. They have two handicaps. One is the latency problem associated with the fact that the speed of light is too slow to avoid perceptible delays in making a 44,000-mile round trip. The other is that uplink equipment remains far too expensive for consumer use. Hybrid systems, like those planned for cable television with one-way cable modems, may be feasible. Hughes is already experimenting with such a service, called DirecPC. Users employ an ordinary modem and telephone line to uplink to the Internet, but the downlink material comes over the DSS satellite on shared capacity. A similar system, CyberStar, is to be launched in 1999. CyberStar plans to offer not just interconnection but content, including multicast videos.

At the other extreme, the first generation of LEOs, including Teledesic, do not seem to be in the running as video delivery vehicles. Even with major improvements in compression, they do not have the capacity. Teledesic is fully interactive (switched), and interactive networks are an inefficient means to deliver broadcast data. GEO Ka-band satellites are also interactive. Even if users are willing to put up with the half-second delay in a GEO-mediated telephone conversation, they are unlikely to put up with it for such mundane activities as channel surfing and video game playing. For these reasons, it does not appear that satellites of any description will become the basis for the delivery of conventional interactive television to the home. They could, of course, continue to be used as conventional direct-to-home broadcast satellites, but such use would waste their interactive capabilities.

Conclusion

Broadcasting television directly to the home from a satellite in geosynchronous orbit has been a dream for many years. Only with the development of Ku-band satellite technology and digital video compression has this become economically feasible.

There are now several competing direct broadcast satellite systems, each offering around 200 channels of television, near video on demand, and audio programming. The programming available on satellite systems is essentially the same that is seen on cable television systems, except that there is more of it.

The chief handicap of DBS service is its present inability to offer local content, such as local TV signals. There is no technical reason why this could not be done; it is a matter of economics. Rupurt Murdoch tried and failed to put together a project that would offer local television signals in major cities.

While the present emphasis is on GEO satellites, new LEO systems may also hold promise. These low Earth orbit systems have the disadvantage of requiring many satellites, not one. They have the advantage of lacking the transmission delay of GEO satellites, and therefore lending themselves to two-way service. Current LEO systems are designed to offer mobile telephone service; the more ambitious ones such as Teledesic and Celestri expect to have the capacity to provide Internet access at speeds comparable to present ones. While no planned LEO system has the capacity to offer mass market video services, advances in compression or storage may permit this in the future.

CHAPTER FOURTEEN

WIRELESS CABLE

Thus far we have focused on the three principal media for delivering video to the home: conventional TV stations, cable television, and broadcast satellites. Although only the third is currently a digital medium, the others are in the process of digital transformation. There are three further contenders for the role of digital television medium of the future: telephone wires, "wireless" cable, and storage media. The various technologies by which television could be delivered via telephone wires were discussed in Chapter 10. In this chapter we survey one of the least-understood technologies: gigahertz terrestrial broadcasting, sometimes called wireless cable.

The term "wireless cable" came into use in connection with a not-very-successful analog service called MMDS, or multichannel multipoint distribution system. Two newer services not yet in wide use in the United States for broadcasting, local multichannel distribution service (LMDS) and 38-gigahertz, along with digital MMDS, perform essentially the same function as MMDS, and all are encompassed here by the term wireless cable. Figures 14.1 and 14.2 illustrate the operation of digital MMDS and LMDS.

The wireless cable business is divided into industry segments, each segment defined by its FCC spectrum frequency assignments. (That is to say, even though the frequencies are auctioned, the segments are defined by the FCC, and therefore are probably quite unlike anything the market would produce.) All of the segments offer television service by similar means. They all have highly localized broadcast areas (from a few blocks to five or ten miles), relatively low broadcast power, very high channel capacity, and the potential for two-way broadband service. Also, each of these segments suffers to some degree from the same problem that afflicts digital satellite broadcasts: attenuation or loss of the signal from heavy rain or dense foliage. Especially at the higher frequencies the receiving antenna must be in view of the transmitting antenna. To a much greater degree than ordinary radio and TV signals, wireless cable signals do not pass through most solid objects. Unlike GEOs, however, wireless cable does not have a latency problem or a local content problem.

From a conceptual point of view, wireless cable broadcasting and ordinary broadcasting are alike: a transmitting antenna emits signals that are picked up by receiving antennas; the strength of the signal, the antenna height, and the sensitivity of the receiving equipment determine the size of the reception area. All broadcast transmissions, however, differ in a systematic way according to the frequency band they use. TV channels 2 to 13 use a total bandwidth of 72 megahertz in the frequencies from 54 to 72, 76 to 88, and 174 to 216 megahertz. Adjacent channels cannot be used in the same area by different broadcasters. UHF television uses a total bandwidth of 336 megahertz in the frequencies from 470 to 806 megahertz with the same caveat. MMDS uses 200 megahertz of bandwidth in the 2,500 megahertz band, and LMDS uses 1,150 megahertz in the 28,000 megahertz band. The higher the frequency, the greater the extent to which the signal is absorbed by the atmosphere, especially by precipitation. On the other hand, higher frequencies are less crowded with competing uses. The new LMDS licensees, with their 1,150 megahertz of bandwidth, have more than all TV and radio stations and cell phone operations com-

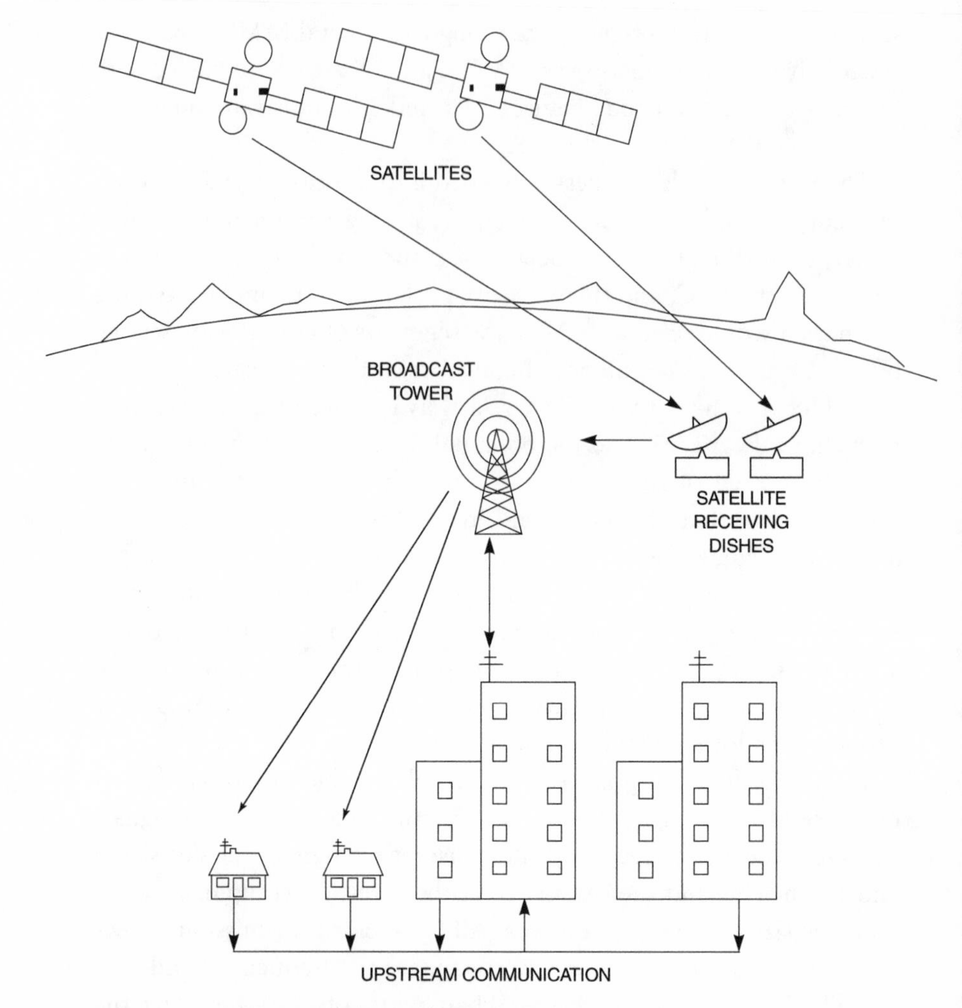

Figure 14.1 Operation of a digital MMDS "wireless cable" system

bined. Not coincidentally, this is approximately the same bandwidth as the most modern hybrid fiber-coaxial cable systems. Up to five 38-gigahertz licensees will share 3,000 megahertz of bandwidth in each city, the equivalent (before spectrum reuse) of three modern high-capacity cable television systems. Figure 14.3 compares the bandwidths of current and proposed wireless broadcast systems. (An ordinary telephone wire has 4 kilohertz; a cellular telephone system uses

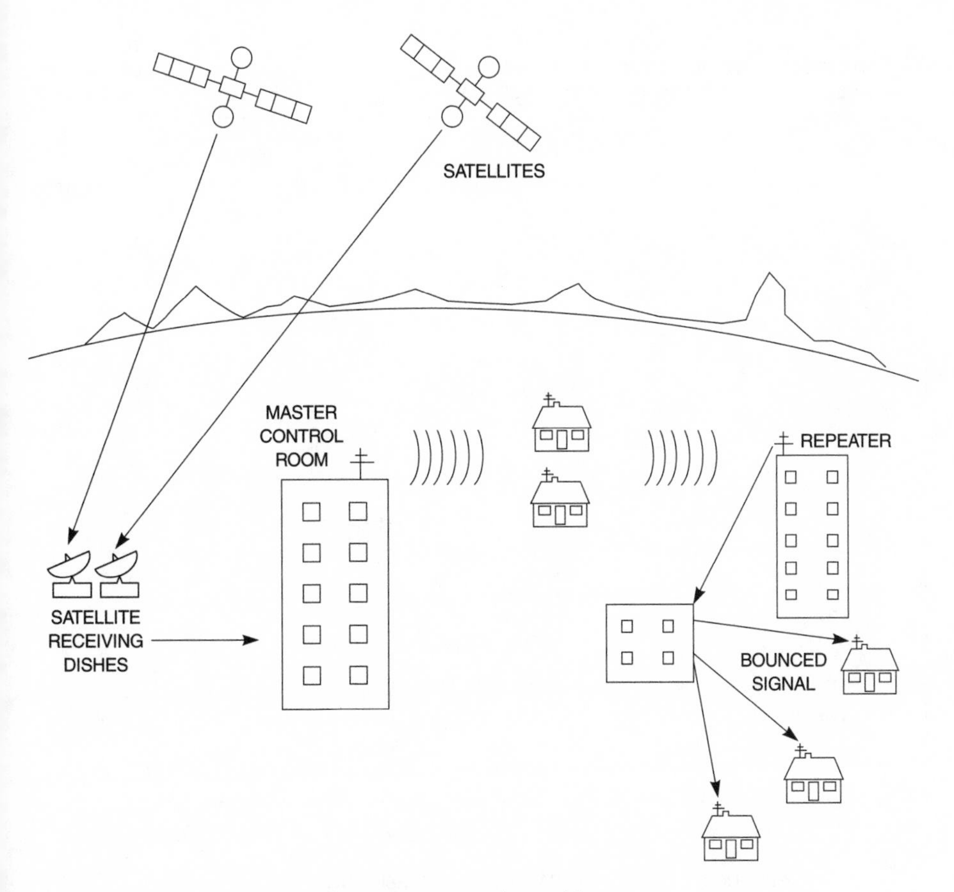

Figure 14.2 Operation of an LMDS "wireless cable" system

25 megahertz per city (before frequency reuse), and a cable system has up to 1 gigahertz of bandwidth.)

The ability of wireless cable to compete with both cable and ordinary broadcasting depends on whether the demand for additional channels (plus freedom from the FCC's broadcast regulations) offsets the cost of new (and possibly more expensive) equipment. In some ways the experience of UHF television in competing with VHF television is instructive. UHF television stations as a group offered viewers additional choices. But in the early years few viewers were willing to pay the cost of the additional equipment needed to receive UHF signals. Even when the FCC required all new TVs to include UHF tuners,

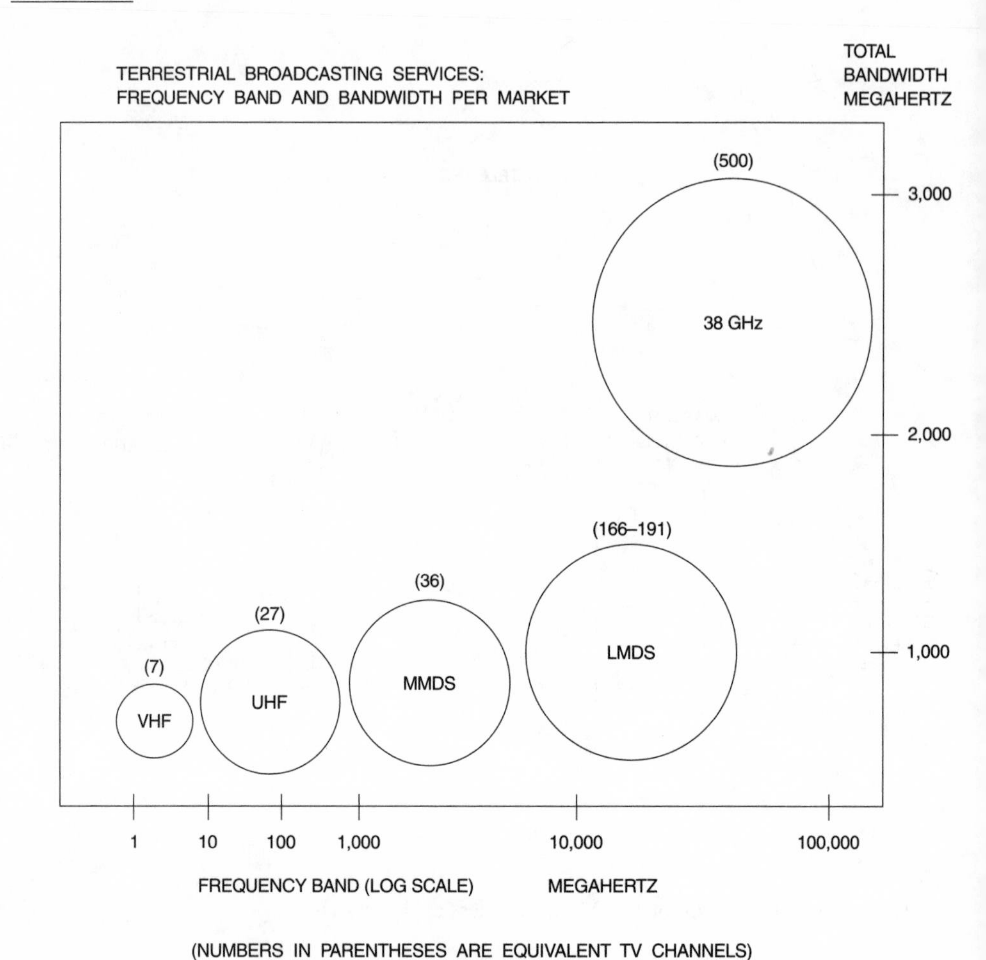

Figure 14.3 Comparisons of available bandwidths for wireless media

many viewers did not take the trouble to explore the UHF options. The situation with wireless cable is somewhat more hopeful because the program content it can provide is already familiar to most viewers and is promoted on other media, especially cable. Unlike UHF television, wireless cable is, or can be made, interactive, enhancing the range of services that can be provided.

The ability of wireless cable to compete with broadcast satellites and cable television systems is an important unknown affecting the future of television. Like cable, wireless can offer local content under a compulsory license; satellites can do so only at substantially higher cost. Beyond that, all three wireless cable media have very high channel capacity. The differences among them in potential capacity are probably not significant, given the role of digital compression. (On the other hand, conventional broadcasting, even when digital, has significantly less capacity.) All three media can offer essentially the same content. All three are, or can be made, interactive at least in the sense that near video on demand could be provided. They can also be made fully broadband interactive if the demand exists to support the expensive consumer-level equipment. The key factors appear to be cost and timing. The technology that is cheapest per channel per household stands a good chance of winning the race, unless consumers have already committed themselves to a medium that got there first. In this respect, of course, wireless cable is at a significant disadvantage; others are indeed there first.

Firstborn: MMDS

The firstborn of the wireless cable services, multichannel multipoint distribution systems, was described in Chapter 8 in connection with the C-band satellites that enabled MMDS to coexist with cable systems by feeding the same programming to both. The typical MMDS system today, however, has only about half the channels of an average cable system. It is an analog service with limited capacity. MMDS is used chiefly to supply programming to multiunit dwellings and office buildings where cable service is not available, more expensive to install, or where the building owner wishes to supply TV programming to tenants at a lower cost than can be obtained from the local cable operator. A few of the large regional Bell telephone companies have toyed with MMDS as a means to compete with cable. Pacific Bell and Bell South did buy, and Bell Atlantic almost bought, large MMDS companies. As it has become more apparent that cable companies are not immediate competitive threats in the telephone business, telephone company interest in video competition, via MMDS or otherwise, has waned. Apparently telephone companies had little interest in offering video service as a separate line of business. Instead, they saw video investments chiefly as counters to anticipated cable entry into the telephone business.

SBC Communications, which inherited the Pacific Bell Video Services MMDS systems, is reported to be pessimistic about the prospects for its substantial and expensive new MMDS system in Los Angeles. The 150-channel digital system, called Pacific Bell Digital TV, apparently cost SBC and its predecessors around $300 million to develop. The system covers most of Los Angeles and Orange counties from antennas on Mt. Wilson and Mt. Majeska. (The Orange County system is analog. PacBell has installed similar systems in San Diego, the Bay Area, and other heavily populated areas of California.) Its content includes local TV stations and the usual cable networks. The local stations give it an advantage over otherwise comparable DirecTV content, and the absence of a hard-wired infrastructure gives it an advantage over digital cable upgrades. On the other hand, its bandwidth places it at a potential disadvantage. MMDS operates with a bandwidth of 200 megahertz, LMDS and advanced cable systems at 1,000 megahertz, and satellites at 500 megahertz. Although MMDS does not face the cost of building a wired backbone, its relatively small scale increases the cost of customer-premises wiring and equipment. Estimates of per-subscriber capital costs for digital MMDS service run as high as $1,700 at expected penetration levels.

Although SBC Communications and Bell South have held on to their MMDS investments, at least for now, other telephone companies have not. Nynex and Bell Atlantic backed off from a $100-million investment in CAI Wireless, an experimental two-way video and Internet offering in several East Coast cities. Among the reasons cited was CAI's ability to reach only 75 percent of the population from its 400-foot antennas. (There are no convenient mountains in Boston or Virginia Beach.) Repeater transmitters or "beam benders" can be used to reach areas shadowed by terrain or buildings, but it is tougher to deal with foliage.

Early in 1996, when telephone company interest in MMDS was at its peak, the FCC auctioned off a set of "overlay" MMDS licenses, generating aggregate revenue of about $220 million, or 86 cents per capita for the covered population. These numbers do not suggest that anyone has high hopes for MMDS.

Times Ten: LMDS

LMDS (local multipoint distribution service) operates at 28 and 31 gigahertz, a frequency band (the "Ka-band") shared with GEOs and LEOs. Under current FCC rules, there are two LMDS licensees in each city, determined by an auction that was held early in 1998, generating $580 million for the Treasury. The bandwidth available to the larger of the two is enormous: 1,150 megahertz, the equivalent of 167 ordinary analog TV channels, which can be multiplied in the usual way by digital compression. Even this understates the capacity of LMDS systems, because their frequencies can be reused by means of small cells and low-power transmitters, the equivalent of spot beams on a satellite or of cells in a cellular phone system. In addition to the 1,150-megahertz licensee, each city also has a smaller 150-megahertz allocation for an independent licensee.

LMDS is a cellular service of necessity: at 28 gigahertz, signals are quickly attenuated by terrain and atmospheric moisture. But a cellular structure is also an advantage because it multiplies capacity through spectrum reuse. LMDS also permits "bandwidth on demand." To provide service to a new customer in the coverage area, only subscriber-end equipment need be installed. If the system is interactive in the video-on-demand sense, bandwidth can be allocated to individual users in a cell as needed, yielding savings through time sharing of frequencies; there are no dedicated user channels, as with telephone-based systems. In comparison with local telephone loops and cable drop lines, this provides the opportunity for substantial economies. Finally, the technology may permit an interactive service in the sense of video telephony or broadband Internet links; this requires more sophisticated and expensive transmitting equipment for each subscriber. According to one estimate from Hewlett Packard, LMDS might be able to produce 7 megabits per second downstream and 1 megabit per second upstream simultaneously per household. Another LMDS firm, CellularVision, has been testing a wireless modem for Internet connections that it claims will offer data rates up to 54 megabits per second, far more than most personal computers can handle. Figure 14.4 shows an LMDS system configured to deliver Internet content as well as television signals, with the telephone system used for upstream traffic.

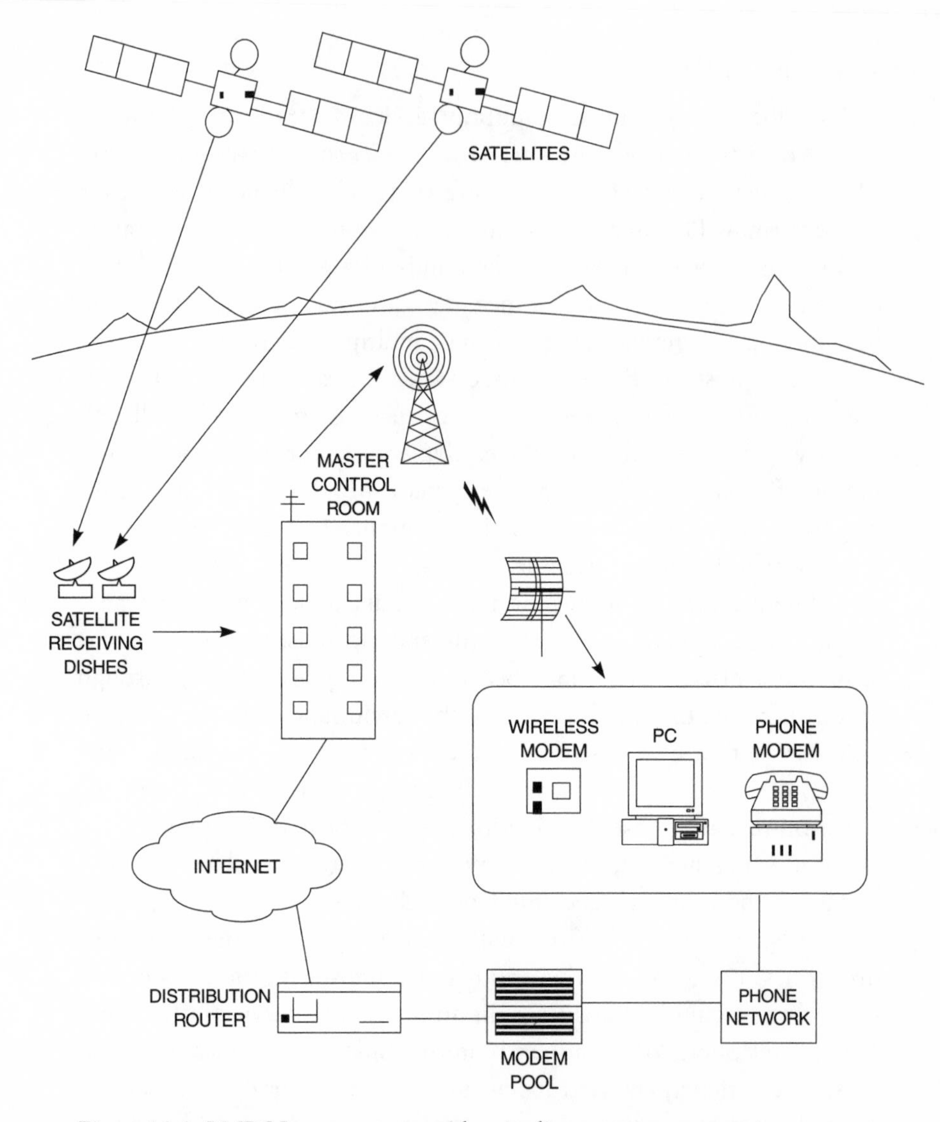

Figure 14.4 LMDS Internet access with voice line upstream connection

This network architecture is essentially the same as that proposed for one-way cable modems.

An experimental LMDS license, the first, was awarded to CellularVision in 1991. CellularVision built and operated an analog LMDS system in portions of New York City, offering one-way video service. A number of technical and other obstacles had to be overcome, and

CellularVision became enmeshed in a protracted struggle with satellite companies for Ka-band spectrum allocations. CellularVision avoids the cost of local trunk distribution to cell sites by "broadcasting" a signal at above-rooftop levels; this signal is picked up and rebroadcast locally at cell sites. Receiving antennas are small, flat objects that can be mounted almost anywhere on a building, or even on the inside of a window. Other experimenters have opted for smaller cell sizes (down to a radius of two kilometers). However, no one really doubts that the technology is viable and that, at scale, it has a decent chance of coming out at a lower unit cost than other broadcast or interactive digital video transmission media. Because of this potential for competition, the FCC excluded telephone and cable companies from bidding for LMDS licenses.

No Acronym: 38-Gigahertz Band

The third segment of the wireless cable market has no name except "38-gigahertz band." At that rarefied "millimeter" frequency band, several companies, but chiefly Advanced Radio Telecomm (ART) and WinStar, have acquired FCC licenses covering most of the populated areas in the United States and have developed an extremely high capacity interactive service that is already on the air in several parts of the country. Meanwhile the FCC has proposed to double the amount of spectrum allocated to this use and to auction it to as many as five companies in each geographic area.

The 38-gigahertz band is even more susceptible to transmission problems from moisture than the 28-gigahertz LMDS band. Consequently, cell size must be smaller (one to three miles of unobstructed line of sight, depending on climate). In residential areas aesthetic problems sometimes arise because of the required antenna height and the number of antennas. Despite these drawbacks, 38 gigahertz is now used successfully in Europe and elsewhere for cell phone service, and it is used for video in Venezuela and Canada.

WinStar and ART claim to be able to link office computer networks, cell phone switches, and other concentrations of traffic with one another or with the Internet at capacities up to 45 megabits per second, with higher speeds to come (see Figure 14.5). This is far beyond anything that can be achieved at present over ordinary telephone

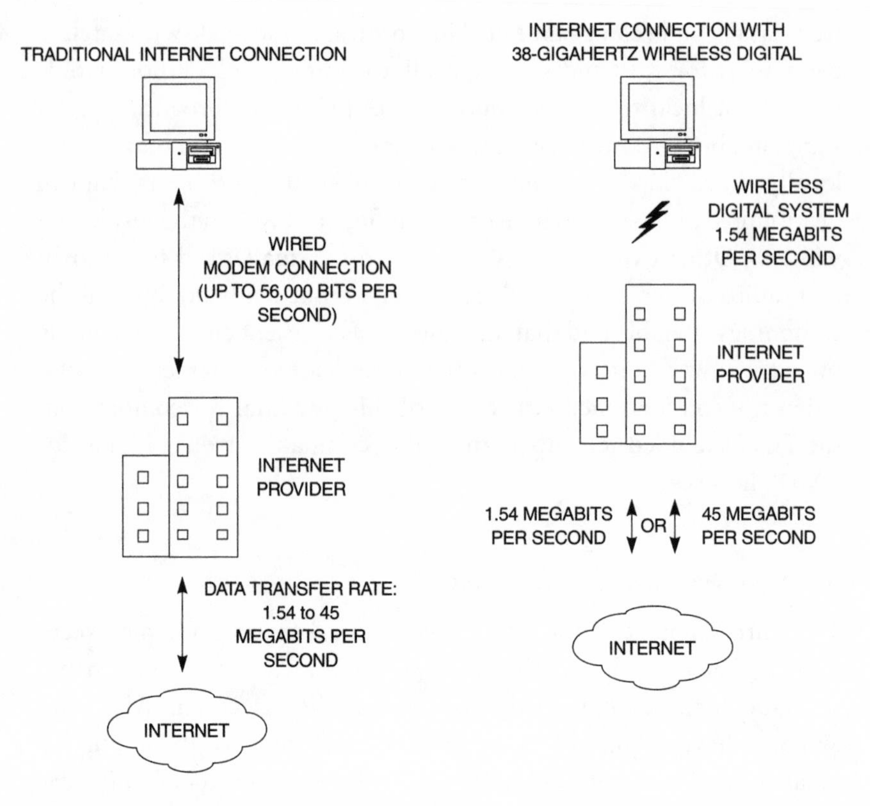

Figure 14.5 Two-way Internet communication using 38-gigahertz wireless connection

lines or cable modems, and even more than the ambitious Teledesic LEO system plans to offer. ART has entered into a partnership with the huge midwestern telephone company, Ameritech, to provide wireless broadband services in that territory.

Both ART and WinStar have positioned their services initially as a substitute for high-capacity local phone lines, rather than as a broadcast video service. One reason for this is that the terminal and antenna-related equipment that would be required for each household is too expensive at present for a consumer service. However, the 38-gigahertz band has a clear potential to offer "last mile" broadband connections to the home, with the prospect of interactive (on-demand) video, videophone, and other similar uses. Indeed, one can think of 38 gigahertz as a kind of laser beam carried by the airwaves rather than by a fiber-optic cable. Broadcast service via 38 gigahertz may consist of separate antenna pairs for each served location. In functional terms it is a matter of indifference whether it is thought of as a broadcast medium or a telephone medium.

Better than Digital Television?

Digital wireless cable is a very promising technology. It has vast capacity compounded by spectrum reuse, low latency, the ability to offer local content, and costs that potentially are lower than those of wired services. Compared with satellites and wired systems, wireless cable requires a small initial capital investment for infrastructure. On the other hand, wireless cable technology is very new, the problems are not yet worked out, and it remains to be seen whether the problem of rain attenuation can be overcome in a way satisfactory to consumers.

It is interesting to compare the new digital channel assignments that will be developed by conventional broadcasters with wireless cable. Even collectively, the conventional broadcasters in most cities will have far less capacity than digital MMDS, LMDS, or 38-gigahertz services. The conventional broadcasters, however, have three advantages. First, their signals are not affected by weather conditions or line-of-sight problems. Second, consumers will be able to receive regular broadcast signals on the new digital TV sets (if such sets are purchased) without a converter box. (Perhaps this will be true for wireless cable as well. The FCC has the authority to require manufacturers to make TV sets that receive 38-gigahertz or other digital signals just as it did for UHF in the 1960s. "Universal" digital TV sets and converter boxes will be a contentious policy debate.) Third, broadcasters, on account of their obligation to continue to provide at least one channel of "free TV," probably can look to the government for protection from a too-successful wireless cable industry.

Conclusion

Wireless cable systems—MMDS, LMDS, and 38 gigahertz—have become available in several different frequency bands. With digital compression, each of these technologies can carry hundreds of television channels as well as one-way or even (at greater expense) two-way

Internet data transmission. Wireless cable systems have channel capacity comparable to that of cable television and GEO satellites. They have the advantages of not requiring connecting wires or cables, and of being able to offer local content. They have the potential disadvantage of being subject to outages during heavy rainfall.

Wireless cable systems appear to have substantial capacity advantages over the new digital television technology as defined by the FCC.

CHAPTER FIFTEEN

DIGITAL TELEVISION

A camel is said to be a horse designed by a committee. A wildebeest is an impala designed by government regulators: ugly, slow, and mindless, with a strong herd instinct. Nothing better fits this metaphor than the Federal Communications Commission's approach to the establishment of digital television standards.

Reality and appearances are not always the same. In an April 7, 1997, *New Yorker* article, Malcolm Gladwell commended the FCC's role in high-definition television. Comparing the process that led in 1996–1997 to the HDTV standards to the process by which Silicon Valley entrepreneurs bring innovations to market, Gladwell pointed out that both the FCC and the entrepreneurs rely on heavy involvement by customers in product development. Few successful software products are shipped without extensive consumer testing and feedback, which often changes the nature of the product. According to

Gladwell, the FCC acted as a proxy "customer" in the development of HDTV standards, poking and prodding the industry in desired directions.

Wildebeest Design

There is a major hole in Gladwell's argument. The FCC is not the customer; it is not in the position of the Tennessee Valley Authority, buying a hydroelectric dam. The government here stands between the supplier and customer, dictating outcomes to each. The result was that digital television standards were adopted, and enormous investments are being forced on the broadcasting industry, without any serious testing by customers, test marketing, or other attention to the demand side of the market. As a result, digital television, or at least HDTV, has all the earmarks of a disaster, and not even the FCC's last-minute decision to require broadcasters to convert to digital transmission but not necessarily to HDTV may save the day: Congress seems prepared to insist on HDTV.

Conventional television broadcasters were cast into the digital age in 1997, when the FCC adopted a plan (later modified by Congress) calling for the phase-out of analog broadcasting by the end of the year 2006. Under the FCC plan, each conventional broadcaster has been granted a new 6-megahertz spectrum assignment to be used for digital broadcasting.

According to the industry committee that defined the new digital standards, the system

> is designed to transmit high quality video and audio and ancillary data over a single 6 megahertz channel. The system can deliver reliably about 19 megabits per second of throughput in a 6 megahertz terrestrial broadcasting channel and about 38 megabits per second of throughput in a 6 megahertz cable television channel. This means that encoding a video source [i.e., HDTV] whose resolution can be as high as five times that of conventional television (NTSC) resolution requires a bit rate reduction by a factor of 50 or higher. To achieve this bit rate reduction, the system is designed to be efficient in utilizing available channel capacity by exploiting complex video and audio compression technology.

During a transition period, according to the FCC plan, both analog and digital broadcasts will take place. After 2006, supposedly all analog channels must be returned to the government, which will already have auctioned them off for other uses. However, Congress in Title III of the Balanced Budget Act of 1997 granted analog broadcasters the right to keep their current licenses (as well as the new digital ones) as long as 15 percent or more of the TV households in any given market have not purchased or leased digital receiving equipment. As Figure 12.2 suggests, few consumer innovations are so successful that they reach 85 percent penetration in a decade. It is extremely doubtful that digital television will achieve this level. It is likely that 15 percent or more of TV households will fail to meet this criterion for many years, permitting broadcasters to retain two channels indefinitely. Quite possibly the FCC will be obliged to exercise its power over TV set manufacturers to require them to incorporate digital tuners in analog sets before the goal can be met.

If digital television actually comes to pass, consumers in markets where 85 percent of the households can receive digital signals, and who wish to receive over-the-air TV broadcasts after 2007, will have to purchase either a new digital TV set or a $400–$600 converter box to translate the new signals into a form their old sets can understand. Broadcasters will need to invest in expensive new transmission equipment, and probably new studio equipment as well. (Each station will need a new tower plus $1 million in new transmission equipment.) These costs will be greater if digital TV takes the form of HDTV.

Digital television (DTV) has been around for some time. For several years, DirecTV has broadcast digital pictures from space using the same MPEG2 encoding standard that has been adopted for DTV. Regular TV broadcasts and cable channels are both digitized for part of their journey from the studio to the TV set. These developments have taken place naturally, because of the superiority or economy of digital transmission. In a free market environment, digital transmission and fully digital reception equipment for consumers would no doubt already have replaced analog technology.

Broadcasting is not, however, a free market. For more than seventy years the broadcasting industry has been an extension of government. Over-the-air broadcasters could not begin using digital technology until the FCC had orchestrated an elaborate, lengthy, and intricate dance of industry titans. Only when "consensus" arose among the various interest groups could the digital conversion begin. The focus of the dance was the adoption by the FCC of a set of technical standards to which all broadcasters must by law conform. While certain aspects of the standards have been left flexible, the basic framework for digital broadcasting is carved in legal granite.

Despite the specific transition plan, considerable uncertainty about the future of digital broadcasting remains. From a technological and regulatory point of view, there are a number of options. One option is HDTV, sometimes called ATV (advanced television). Originally, the FCC used "high-definition television" (HDTV) as an umbrella term to describe the new service. HDTV was to have higher-resolution pictures and better sound than conventional television. During the long process of negotiating a standard for the new service, digital television was developed. The FCC then adopted the term "advanced television" (ATV) to refer to the more flexible digital service. Subsequently, the commission adopted the term "digital television" (DTV) to refer to the service; HDTV is one subset of DTV.

HDTV pictures will have the same rectangular or "letterbox" shape as a movie theater screen, and the quality of the signal will be comparable to that of a movie. Current television pictures are too grainy to show that the anchor desk on the six o'clock news is made of cardboard, but HDTV pictures will show it all. TV studio sets, often crudely constructed and decorated, will have to be rebuilt or replaced with computer-generated "virtual" sets. Even more difficult to hide will be facial imperfections; ordinary makeup will no longer do the trick. On the plus side, high-definition digital video will be accompanied by six-channel Dolby surround-sound.

If a broadcast station chooses to offer HDTV pictures, it will be able to offer only one, or perhaps two, channels using its new spectrum assignment. But there are alternatives. The same broadcast station may choose to devote its 6-megahertz digital assignment to broadcasting pictures of present (or even lower) quality. In that case, six or more simultaneous programs could be broadcast, and viewers might be asked to pay for some of them. The term "multicasting" has emerged to describe this strategy. Finally, the broadcast station may choose to use its 6 megahertz to broadcast one or more signals of current or slightly better quality and to devote the rest of its capacity to some other use, such as paging or wireless Internet connections.

Under the FCC's digital standards, a TV station can broadcast in any of eighteen formats. Only a few of these are immediately relevant: those with progressive (P) or interlaced (I) scan lines, and with 480, 720, or 1,080 lines per frame. The highest resolution television picture that can be squeezed into a 6-megahertz TV channel with current compression technology is 1,080I. This is referred to as high-definition television, or HDTV. (1,080P would be of higher quality but is not yet attainable.) Some experts, however, claim that 720P offers picture quality indistinguishable from 1,080I. The new "standard" digital television picture will be 480P, which offers pictures superior to those provided by present-day NTSC analog signals. As of mid-1998 there was no agreement among broadcasters on which format would be used for HDTV. For example, ABC and EchoStar announced plans to use 720P for HDTV, while CBS, NBC, and DirecTV planned to offer 1,080I, and Fox 480P. TCI had not decided whether to offer HDTV at all. For "standard" digital broadcasts, ABC, CBS, and NBC planned to use 480P, while EchoStar, DirecTV, and TCI proposed to offer 480I.

Not all the rules are yet in place, but the FCC seems inclined to permit broadcasters to use their digital assignments for any purpose, as long as they continue to broadcast at least one free TV channel of standard quality. ("Standard" here refers to the new digital standard, so the quality of the picture will be better than present analog quality.) Depending on the economics, there may never be widespread HDTV broadcasts, unless Congress insists. HDTV broadcasts receive the hype, and have been the focus, of virtually every popular media piece on digital television. Congress, hypnotized by the avant-garde technology of HDTV, has been led to grant broadcasters a whole set of new channels. But HDTV's future is nevertheless uncertain. HDTV broadcasting is very expensive both in terms of broadcast equipment and in terms of spectrum use. Moreover, it puts out a signal that is not the passive, low-resolution picture for which viewers have demonstrated a demand. Computer monitors have to be high resolution, but do TV sets? A big screen and sharp details invite heads to turn and eyes to focus, making it harder to carry on a simultaneous conversation with one's sofa mate. Despite a decade or more of work on the standard, no one seems to have made public any serious study of the extent and nature of consumer demand for these services. No doubt individual companies have made private studies. But these have not informed the official decision-making process.

Perhaps the biggest drawback is the hefty price tag. HDTV makes sense chiefly for large-screen TV sets; on small sets the increase in picture quality is far less noticeable. Manufacturers have announced that early HDTV sets will cost $8,000–$10,000. To receive HDTV signals as such requires consumers to purchase new TV sets; converter boxes that translate digits into analog wave forms will not do. (Such boxes can, however, translate HDTV signals into a form that can be displayed on a conventional TV set at a quality slightly better than current analog television.) At this point it is an open question whether consumer demand for the HDTV picture is strong enough to force advertiser-supported broadcasters to use it, despite the higher cost. In any event, it is not likely to arrive quickly: consumers must first purchase a critical mass of HDTV sets, and program producers must produce a critical mass of HDTV content. The rerun programming that makes up the majority of television fare today, while it can be converted to appear on an HDTV screen, will not look very good because it was not produced with HDTV in mind.

Broadcasters who choose to offer multiple channels of present-day-quality digital programming will face other difficulties. First, there is no assurance that these new channels will be carried by cable or satellite operators. Today, a law requires cable operators to carry all local TV signals. Some believe that this law will be extended to satellite companies that carry any local broadcast signals. That would greatly strain an already weak rationale for the law. Why insist that multiple broadcast channels—and probably nonfree channels at that—be carried by cable or satellite operators?

Second, multiple new channels are more likely to survive if subscriber support is added to advertiser support. But monitoring viewing and collecting payments are very expensive on a per-channel basis when so few channels are involved. Nothing distinguishes a digital broadcaster using 6 megahertz to offer six or eight digital TV channels from an LMDS or MMDS operator, except that the latter has far more channels and roughly the same distribution costs, spectrum aside. Handicapped by the requirement to continue broadcasting at least one free channel, broadcasters may not survive such competition.

One obvious solution would be for broadcasters to pool their spectrum resources (beyond the amounts necessary for each to continue to provide a channel of free programming). The pooled resources in most big cities would be adequate to produce an over-the-air multichannel competitor only marginally able, in terms of capacity, to vie with wired media or other wireless media. (The average viewer receives twelve over-the-air signals; the average maximum joint broadcaster capacity is therefore about seventy-two digital channels.) There would not be much to distinguish it from an MMDS system. Further, such a combination would not have a distribution cost advantage because the government expects to charge for spectrum put to such uses, unlike free over-the-air broadcasts.

The Rules of the Game: Standards

Megahertz digital broadcasting (that is, broadcasting of digital TV on the existing TV channels) is of interest from another perspective: the development of technical standards. There cannot be a more dramatic contrast in the history of standards than that between digital broadcasting and the Internet. Even though both make use of the same underlying technology (digital transmission), their paths have been strikingly different. The Internet has standards—generally accepted ways of doing things—that are constantly changing as technology and entrepreneurial fortunes change. The pace of innovation on the Internet has been phenomenal. No one with a new idea is required to get permission from the government or even from a private standards organization before offering a new service. Of course, if the new service is not compatible with existing standards, it will have to be very good indeed for users to adopt it. Nevertheless, that happens all the time. World Wide Web sites (servers) were useless without World Wide Web browsers, and vice versa, but the technology spread like wildfire because of its superiority to the File Transfer Protocol (FTP) and other methods of Internet search and data transfer. The programming language Java, for example, has not been officially accepted as a standard by industry organizations, but it is nevertheless in widespread use and may become a de facto standard.

Standards controversies can be resolved privately either by the market (as in the VHS versus Betamax videocassette war) or by voluntary agreement. Two companies, Rockwell and U.S. Robotics, developed and produced incompatible 56k modems. In less than two years they (and other industry participants, including customers) agreed on a compromise voluntary standard embodying elements of both designs.

In contrast, it took a decade of ponderous government-industry consultation for an acceptable digital television standard to emerge and be adopted by the FCC. Any important industry group could use political clout to block implementation of a standard with which it did not agree; all groups had to agree on the new standard. Meanwhile, conventional broadcasters were not free to experiment, because the FCC specifies all the important dimensions of their transmission procedures. Broadcasting any other way is against the law.

Although more flexible than the old standard for conventional television, the new digital television standard is still a set way of doing things, and in the digital age any fixed way of doing things today is likely to be inappropriate tomorrow. Enshrined in regulations, the digital television standard can be changed only with difficulty. The regulations incorporate by reference a private sector document, the Advanced Television Systems Committee Digital Television Standard, which in turn makes reference to various International Standards Organization standards. The need to consult and move these bodies makes the already time-consuming rulemaking process even lengthier.

For these reasons, broadcasters using digital transmission techniques may one day face competitors using superior methods and be unable to adopt the new methods themselves. This is another reason to doubt the long-term economic viability of digital megahertz broadcasting.

One important feature of digital communication is its flexibility. Unlike most analog channels, a bitstream is transparent to message content and form. The digital processor at the receiving end, acting on built-in instructions or instructions in the bitstream itself, constructs the message. Analog methods, at least in practice, lack this flexibility. To adopt technical standards for digital broadcasting as a matter of law is, by definition, to limit the range of possible future television forms. This is not to say that technical standards are bad. Far from it. Standards can promote diffusion of innovations and lower costs, especially when they arise from decentralized decision making by industry participants, but also when they are established by nonbinding agreements among competitors. (Some, but not all, of the movement toward standardization of rail gauges among nineteenth-century American railroads resulted from such agreements, for example.) Partly for good reason, even voluntary or "natural" standards are hard to change. But standards set by government regulation with the force of law are almost impossible to change.

The FCC technical standard for analog television was established in all important respects except for color in the 1940s. When color was added in the early 1950s, the standard was locked in for fifty years. AM/FM radio has a similar story. What other electronics industry has had its technology frozen in place for fifty years? Chances are, absent the FCC lock-in, we would already be enjoying some form of advanced digital television.

The Interlace Controversy

One of the standards issues not settled by the FCC decision on digital television was the convention for scanning. Conventional television pictures are transmitted using "interlace" scanning: every other line of a frame is sent as one field, then the remaining lines as the next field, at 60 fields, or 30 frames, per second. The quality of a television signal is greater, other things being equal, the larger the number of frames per second. If there are too few frames, the retentivity of the human eye (or brain) is not sufficient to invent the intervening motion. It sees flickering images. Other things being equal, the frame rate is limited by bandwidth. In order to maximize the perceived quality of television images confined to a 6-megahertz channel, television engineers invented the interlace approach, in which visual retentivity is harnessed to create a perception that the frame rate is greater than 30 frames per second by sending "half frames" at the rate of 60 per second (see Figure 15.1).

Given current compression technology, many in the broadcast industry believe that a high-definition television transmission is optimized by using 1,080 lines with interlace scanning. There is agreement that if bandwidth were not a constraint, or if compression were more efficient, 1,080 lines with "progressive scanning" would be preferable.

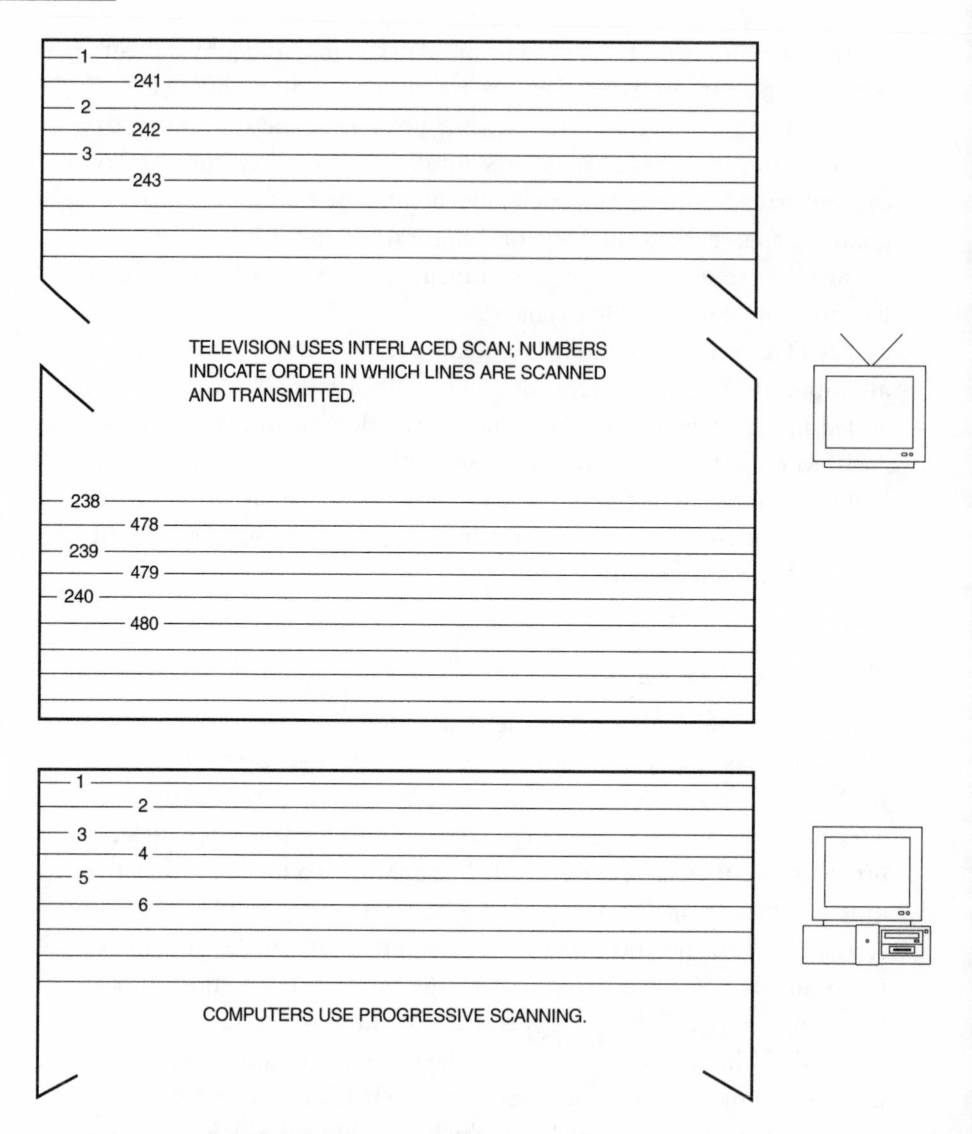

Figure 15.1 Interlaced and progressive scans

With this technique, lines in each frame appear sequentially. Interlace scanning reduces the visual quality (sharpness) of text and other images associated with computer use.

High-resolution computer monitors do indeed use progressive scanning. This is a superior format for the display of text. Unconstrained by bandwidth, computer monitors can display 60 full frames per second, as opposed to 60 fields, or half-frames, per second in a standard TV broadcast.

The computer industry would prefer that the television set manufacturers and the rest of the television industry adopt a progressive scan approach to digital television, but not at 1,080 lines because of the added cost in computer hardware required to display such signals on monitors. With progressive scanning, digital television sets could be used readily and cheaply to display computer or Internet data, and current computer monitors could display digital TV signals. However, to accomplish this the television broadcasters would have to accept a lower-quality "high-definition" picture, with 720 lines rather than 1,080 lines. They would also have to accept more rapid inroads on their market by the computer software and Internet content-producing industries.

To its credit, the FCC has decided that the controversy over interface and progressive scanning is to be settled "by the market." Speeches and notices by leading firms and trade associations in each industry, intermediated by the trade press and the general press, will sway opinion one way or the other. Once again the tradeoff between bandwidth and compression is at the heart of the matter. Everyone agrees that as processor power increases (and as cost per unit of processor capacity falls), progressive scanning at 1,000 lines per frame will become possible. Everyone also agrees that television set manufacturers (and computer monitor manufacturers) could solve the problem by building in circuitry capable of any sort of scanning. They could, but they won't. For now it is too expensive, given highly uncertain forecasts of consumers' willingness to pay for such devices.

Is Digital Television in the Race?

Digital television faces serious handicaps as a successor technology to today's television delivery system. It has the least capacity of any wireless broadband system, even taking the local stations collectively. It is constrained by the political need to continue to provide free TV, and by Congress's desire to see HDTV programs broadcast, whether or not that makes economic sense. In order to achieve anything like an adequate bandwidth and a program package competitive with cable or other wireless media, local stations may have to merge or cooperate, creating more regulatory problems. If digital television were starting from scratch with these handicaps, it is difficult to see how it could survive. But digital television in the real world will start off with some advantages. Chief among them will be a strong viewer franchise and powerful political connections.

Conclusion

Television stations use analog technology defined in 1954. In 1997 the FCC finally adopted a plan for broadcasters to begin using digital technology. Both the old analog standards and the new technical ones constitute rigid constraints on how broadcasters may operate. Innovation is essentially impossible. Because of this, it is likely that digital television would have been in place many years ago but for the FCC's regulations.

Under the FCC's plan, which has been changed repeatedly both by the commission and by Congress, each broadcaster will receive one digital channel, and will give up its old analog channel at some point in the future. Under the standards, using present technology, each broadcaster could use its 6-megahertz digital channel to offer up to five or six "standard" television signals. However, even assuming that all local broadcasters pooled this capacity, there would be fewer channels than what satellites, cable, and wireless cable can offer.

Alternatively, each broadcaster could use the new 6-megahertz channel to offer a single channel of high-definition television. This would be television of theatrical motion picture quality, with advanced multichannel sound. The consumer demand for HDTV is unknown.

Central to the future of television and the Internet is the ability of TV sets to display high-resolution computer images and of computer monitors to display television pictures. The broadcast industry presently uses a bandwidth-conserving technology called interlacing, which is not compatible with standard computer outputs. The outcome of this dispute between the computer industry and the broadcast industry may affect the competitive positions of both equipment and service suppliers in the two industries.

CHAPTER SIXTEEN

TELEVISION BY MAIL

This chapter focuses on digital storage technologies in order to provide some context for the possible futures of television in the digital age. One objective is to describe the storage technologies and how they work; another is to give the reader some sense of what technologies are on the horizon and how they might affect the television industry in the years to come.

Many data storage devices can hold video information: videotapes, Laservision/Selectavision, read-only memory, caches, hard drives, compact disks and CD-ROMs, digital video disks, and even newer optical technologies. Consumers can receive this video information ("television") through the same distribution channels as printed matter—the mail. ("Mail" here is a metaphor for all the physical means used to distribute retail products.)

Storage Media

All media have three things in common, as Chapter 1 explained: channel capacity, storage, and computing power (processing or compression). Efficient media combine these attributes in a way that minimizes cost for any given level of information delivered to the consumer. To be accepted in the market, a medium must be efficient in this sense while satisfying consumer demand more effectively than competing media, at a price that covers its costs.

Storage plays many roles in modern communication media. Print media and sound recordings rely on the manufacture of permanent storage devices that are physically transported to the user. For print, storage is the medium. In contrast, conventional live radio and television use no storage device: their transmissions are entirely ephemeral. In both cases, storage is traded off for bandwidth. The alternative to transport of stored data for print is for potential readers to transport themselves to the publisher's front window, where the news in centuries past was posted. Such transport, a print analog of bandwidth, is clearly less efficient than storage. In the case of radio and TV broadcasting, the opposite is true; transmission is cheaper than manufacturing and physically distributing copies of the material to millions of households. There are two reasons. One is the public-good nature of broadcast transmissions. The other is that the spectrum or bandwidth used by broadcasters has no opportunity cost. Under FCC regulations, bandwidth designated for broadcasting cannot be used for any other purpose.

Just as there are technical means to conserve bandwidth by using computational power to compress data, storage can be used to conserve bandwidth. For example, a memory "cache" can be used to store transmitted information temporarily, until needed, as an alternative to transmitting the same information more than once. A "frame" of a TV signal, for example, may change little or not at all from one-thirtieth of a second to the next. A memory cache can retain the frame, or parts of it, and thus avoid the need to broadcast the same information repeatedly. Caching is done routinely with Internet content accessed through a Web browser. More generally, a memory cache can be used to store information transmitted during periods when the transmission channel is underutilized, thus reducing the bandwidth necessary to serve peak demand. Similarly, a more permanent memory device such as a compact disk read-only memory (CD-ROM) can be used to store images or information. A signal transmitted over a communications network triggers the display of this information on a computer screen or television. The signal requires little bandwidth compared with the volume of information that it may trigger. A specific example is the "interactive character," or IC, developed in the late 1990s as a general-purpose device for bandwidth conservation on the Internet. The idea is for users to install (presumably free) client software that animates stored images of cartoon characters on receipt of a signal from a distant Web site. Web sites can then be designed to send such signals in lieu of some greater number of bits needed for the whole representation or its equivalent.

In each of these examples there is nothing inherently superior about the transmission or storage, or computation capacity. In various combinations, they are just alternative ways of accomplishing the transfer of any given amount of information. Each component comes in a variety of types, qualities, capacities, and prices. Technical progress in each component is rapid and uneven. The combination (medium) that is the most cost-effective means of achieving any goal today is unlikely to be the same tomorrow. At least theoretically, the relative cost of any of these components might suddenly fall by a huge amount, and if that happened the result would be a revolution in media technology.

Where's the Hype?

Anyone researching storage technology encounters an immediate striking fact: compared with the transmission technologies, storage gets hardly any press. Storage is certainly no less important than transmission, if only because cheaper storage means a reduced need for expensive bandwidth. Moreover, technological change in storage has been every bit as exciting over the past several decades as changes in transmission and computation technology. One way to assess the amount of attention paid to a concept in the technical community is to see how many "hits" turn up when a search engine is used to scan the Internet for uses of the term. In the following tabulation the hits reported by two popular search engines (HotBot and Excite) are presented.

Search term	HotBot	Excite
ADSL or ISDN	228,000	130,000
Data storage	109,000	13,900

Why should storage get so much less attention than transmission? It turns out there are simple economic reasons for this disparity. Transmission technologies are usually implemented on a large scale by huge enterprises, and often they are characterized by substantial economies of scale and by network externalities. DirecTV, for example, cost around a billion dollars up front, and each new subscriber is almost pure profit. If the cable industry or the telephone industry implements broadband digital service to residential subscribers, the up-front investment will run into the tens of billions. (Teledesic is slated to cost $8 billion, for example.) And the companies who will do this investing are, for the most part, large.

Storage, in contrast, is a manufacturing business, and one apparently not characterized by overwhelming economies of scale. While manufacturers of memory devices are certainly not all small operations, neither are they the size of AT&T. Moreover, unlike transmission networks and microprocessors, most storage devices do not seem to have the strong network effects characteristic of transmission technologies. Technologies with strong network effects obtain tremendous advantages from being widely adopted at an early stage; this explains Microsoft's vast success with its MS-DOS operating system. Not everyone can be clever or lucky enough to negotiate an exclusive supply contract with the dominant supplier of computers, as Bill Gates did and IBM then was. Hype is the next best thing. If the buzz is that so-and-so's transmission standard or system software is so great that everyone else will have to adopt it or be compatible with it, then it is possible that everyone will, whatever the merits. Competition in Silicon Valley requires as much attention to cultivation of buzz as to cultivation of bits.

Storage is different. A floppy drive or a hard disk drive must conform to certain standards in order to be interchangeable with other such devices and thus operable on a computer designed with that standard in mind. But computer and storage technologies change so fast that technology lock-in is not a serious problem. A firm that decides to go it alone with a new or proprietary interface standard does not face the same harsh penalties for failure that it might if change were less rapid. The same thinking permits customers to purchase new storage products at lower risk. For all these reasons, there is less media hype about storage than transmission systems.

Types of Storage

Digital storage devices usually are evaluated according to three performance characteristics: seek time, data transfer rate, and density. Seek time is the average time it takes to find any given datum stored in the device; transfer rate is the speed at which data can be continuously transferred from the device to, for example, the processor or the display; density is the amount of data that can be stored on a square inch of the medium. Storage devices are also divided into three broad categories: primary, secondary, and tertiary storage or memory.

An example of primary memory is so-called cache memory, built into a processor chip for the purpose of storing intermediate results or frequently used instructions. Such memory has very fast seek times and data transfer rates, but only limited capacity and high per-unit cost. Another example of primary storage is random-access memory (RAM or DRAM) that is used to store data, program instructions, intermediate results, and video display information. Such devices usually take the form of silicon-based chips and typically are located on the same board or bus as the processor. It is important for RAM devices to have extremely fast seek times and data rates, because no processor, however fast, can perform at speed without correspondingly fast memory. Today's commercial RAM chips have seek times on the order of 0.0001 milliseconds, data transfer rates on the order of 100 megabits per second, and capacities of 64 megabits per chip. Memory chip technology has been progressing rapidly, and before long gigabit RAM chips will be commercially available. On the basis of its performance characteristics, nonvolatile silicon-based RAM would be the medium of choice for all data storage. Such memory, however, remains far more costly per unit than other technologies.

There exists a variety of secondary storage technologies. The most common examples are magnetic hard drives. These mechanical devices, found in virtually every modern PC, can store very large amounts of data (up to about 65 gigabytes at present), but seek times and data transfer rates are much slower than for primary storage devices, which do not have moving parts. A modern commercial hard drive is no different in its basic operation than any magnetic medium, including videotape: data are stored in the form of magnetized specks of iron oxide. Such a device today might have 2 gigabytes of capacity, an average data seek time of 50 milliseconds, and a data transfer rate of 2.5 megabytes per second.

The data transfer rate and seek time of a hard disk are functions of the speed at which the disk is spinning, among other things. Disk speed is limited by centrifugal forces that tend to distort the medium. This is more of a problem with larger disks. It is helpful to minimize the mass of the read/write arms that travel over the spinning disks because of the problems caused by inertia. And so on, through a litany of tradeoffs.

Tertiary storage involves much longer seek times and much greater capacity, and it has the lowest per-datum costs. Reels of magnetic tape, removable hard drives, banks or arrays of hard drives, and CD-ROM "jukeboxes" are tertiary devices. Tertiary storage seek times are measured in minutes or hours rather than milliseconds, and the capacity is one terabyte or more. (A terabyte is a trillion bytes.) The largest tertiary storage systems are off-line libraries of magnetic tapes. Although tertiary storage has the lowest per-byte cost, that cost is not proportionately lower than primary memory costs. In other words, the increase in capacity, while it brings lower unit costs, does not reduce unit costs by the same factor as the capacity change.

All three forms of storage—primary, secondary, and tertiary—are in widespread commercial use, and their costs are falling even as their performance characteristics are rising. Figure 16.1 traces the unit cost history over a recent two-year period. Figure 16.2 does the same for seek time. The technologies with the most desirable characteristics (fast seek and high transfer rate) tend to have the lowest capacities and the highest unit costs. The technologies with the lowest unit costs tend to have the slowest seek times and transfer rates. These characteristics are important for purposes of real-time computation or digital signal processing. They are also important for storage used in connection with interactive video services, such as video games and user-controllable playbacks of video material such as prerecorded movies.

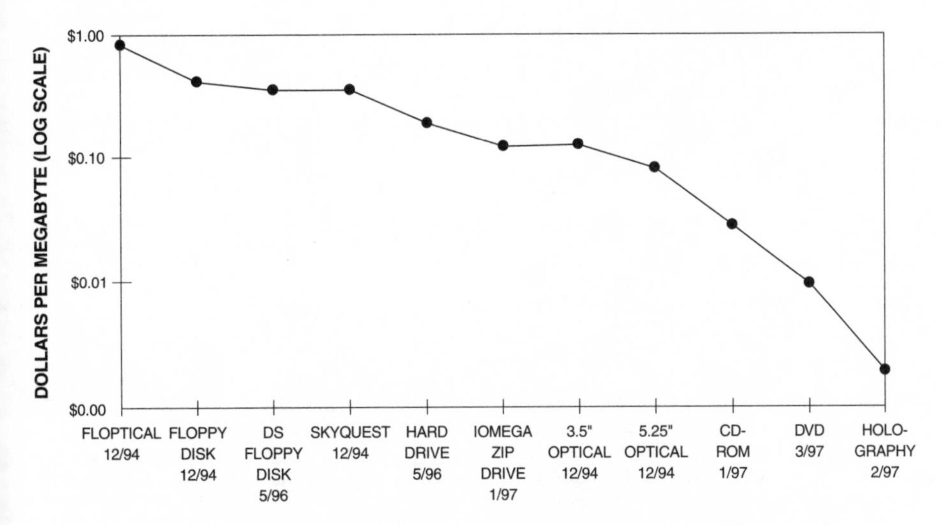

Figure 16.1 Declining cost of storage, 1994–1997 ($ per megabyte) (Schroeder, *PC Week,* Smith, Toshiba Corporation, NCA Computers, 1996–1997)

Although virtually any data storage device can hold video information, some devices, such as present-day floppy disks, have such limited capacity and performance as to make them impractical for video. Today's video storage media include:

- videotape (reel or cassette, VHS or Betamax)
- Laservision/Selectavision (early video disks)
- RAM (cache)
- Hard drives
- Compact disk and read-only memory compact disk (CD and CD-ROM)
- Digital video disk (DVD)

The standard form for consumer storage of video signals is a VHS cassette. There are other tape formats (for example, Beta, SVHS, 8-millimeter); at the production or studio level professional formats are used. All of these videotapes are analog formats, but a digital format would work in much the same way. The basic technology is the same: a magnetic field is modulated by the video input signal, causing revers-

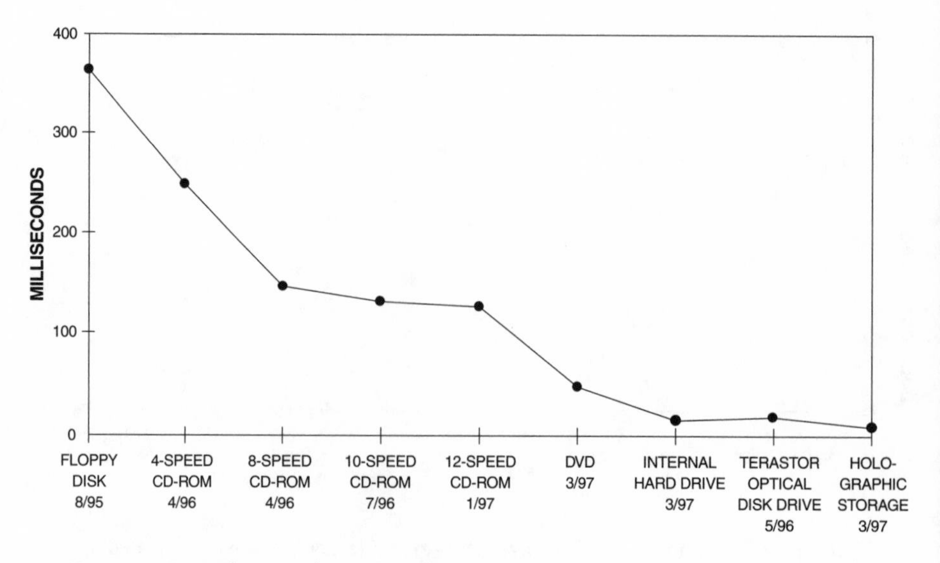

Figure 16.2 Seek time of various storage media, 1995–1997 (Schroeder, *PC Week*, Easystor, Aztech Company, Digiconcepts Company, Toshiba Corporation, 1996–1997)

ible magnetization of small particles of iron oxide that are stuck on the plastic tape. When the tape is played, a pick-up head "reads" the magnetic pattern and converts it back into a video output signal. The quality of the signal depends on the speed and width of the tape (the more square inches of iron oxide that are devoted to a given amount of information, the greater the accuracy of the playback). Videocassettes are an inexpensive and effective way to deliver off-line or nonreal-time video programming. They can be delivered through the same distribution channels as printed matter or groceries. Production costs for prerecorded 90-minute movie cassettes run about $3 per unit in volume. Currently, around 3 billion cassette rentals take place each year in the United States, a number that is declining, while sales of cassettes are rising.

Laservision/Selectavision/Laser Disc

Around 1980, just as the videocassette recorder was gaining a foothold in the consumer market, three video disk technologies were introduced by Philips, JVC, and RCA. All failed in the marketplace, no doubt in part because, unlike tapes, disks could not be used to make home recordings. That cannot be the whole explanation, however, because phonograph disks and CDs have had the same property and have prospered for many years despite the availability of audio recording tape. Perhaps another reason was simply that the new disks were introduced a year or so too late, after VCRs had begun to catch on. The RCA and JVC technologies also had the drawback of being mechanical: a stylus was dragged along a spiral track. The technology was similar to that used in old phonograph records: the signal, imprinted on tiny grooves on a rapidly spinning disk, was read by a pickup head that moved along the groove. These electromechanical devices turned out to have performance features and costs inferior to videocassette tapes. The Philips device, however, was quite advanced, based on laser technology, and differed from modern CDs and DVDs chiefly in being analog rather than digital. The Laser Disc technology, introduced about 1985 by several Japanese consumer electronics firms, has held on since that time as a niche product for videophiles and certain industrial users. The Laser Disc has never reached 2 percent of U.S. television households.

Silicon Devices

Random-access memory chips are an exceptionally fast high-performance form of digital memory used chiefly as part of computer processing. RAM chips certainly can and do hold video. In fact, a computer display is a mapping of the current contents of a portion of the computer's RAM. RAM capacity, however, is normally insufficient to store significant amounts of video information, and RAM is not designed for permanent storage: the information is lost when the power is turned off. Other silicon-based storage devices, such as EEPROMs, can hold data indefinitely even in the absence of electrical power, but they too suffer from capacity limitations and very high per-datum costs relative to other video storage media. Still, there is no reason in principle why semiconductor-based data storage could not be used to store and distribute video signals; as costs decline and capacities increase, that may come to pass.

Magnetic (Hard) Drives

Digital hard drives—a stack of rapidly spinning thin disks, about 3 to 5 inches in diameter, coated with iron oxide—are an integral part of personal computers. Introduced commercially in the 1980s, they supplement the more portable but less capacious floppy disk. They continue to grow in popularity: it is anticipated that around 170 million fixed hard drives will be shipped in 1999, totaling 1.4 million terabytes of storage capacity. Historically, disk drive technology has obeyed Moore's Law: the capacity one can buy for a dollar has about doubled every two years.

Hard drives hold much more data and perform much more rapidly than floppies. Although hard disks are much slower in access time and transfer rates than RAM, they have higher data density. If current trends continue, most hard drives shipped in the year 2000 will have 100 gigabytes or more of capacity, with densities in the range of 10 gigabits per square inch of surface area. Data density on hard drives has been increasing at the rate of about 60 percent per year. In some applications, such as digital cameras, hard drives may replace RAM as on-board storage. Seek time for today's commonplace hard drives is on the order of 10 to 40 milliseconds, and data transfer rates are around 50 megabits per second. On the leading edge are devices like the Travelstar 8GS, a miniature magneto-resistive hard drive with a data density of 3 terabytes per square inch and a capacity of 8.1 terabytes.

An impediment to data transfer rates from hard drives and other storage devices to processors is often found in the speed (bandwidth) of the connecting ports and "bus." Fast processors and fast memory accomplish nothing unless the pipe between them is big enough to handle the flow, and while this creates no theoretical bottleneck, it often creates a practical one.

Compact Disks

Digital compact disks were originally developed by consumer electronics manufacturers for the music industry. Music CDs are more durable and of higher fidelity than audio tape cassettes. Because CDs are digital, they can be used to store any digital information, including material that can be processed by computers. CD-ROM (compact disk read-only memory) simply means a compact disk that holds permanently embedded computer data, including but not limited to video clips. More commonly, CD-ROMs are used for elaborate computer games and storage of materials such as electronic dictionaries and encyclopedias. Compact disks that can be written to, or erased, either once or repeatedly, also exist. All CDs share a few common formats. As a result, any CD player or reader can read any CD. CD-ROM disks can hold up to about 650 megabytes of data. Seek times and transfer rates depend on player technology that varies widely and is constantly improving. Early CD-ROMs transferred data at 150 kilobits per second; current so-called 24X drives transfer information at the rate of 3.6 megabits per second. Seek times are on the order of 150 milliseconds. Speedier drives are introduced each year.

The advent of CD-ROM technology created the opportunity for "multimedia" content. CD-ROMs have sufficient storage capacity to contain a large number of images (or short video clips) and many minutes of recorded sounds. A personal computer under the control of an appropriate program can "play" these images and sounds. This is what is generally meant by the term "multimedia." Multimedia in this sense is not television, and neither CD-ROMs nor personal computers have any native ability to offer television, although a television tuner can be installed in the same box as a computer, can take advantage of some common electronic components, and can use the same display.

CD technology is analogous to the technique used to make old-fashioned phonograph records. One side of the plastic surface of a 12-centimeter disk is imprinted with spiral grooves that have microscopic pits. The pits are created by a laser that vaporizes small amounts of plastic, and the disk is read by another laser that detects the presence or absence of pits corresponding to bits of information (see Figure 16.3). Lands are the spaces between the spiral grooves of pits. Unlike a photograph record, the read head on a CD does not touch the surface of the disk, so there is nothing on the disk that can wear out. Prerecorded CD-ROM disks cost less than $1 each to produce in volume, exclusive of content.

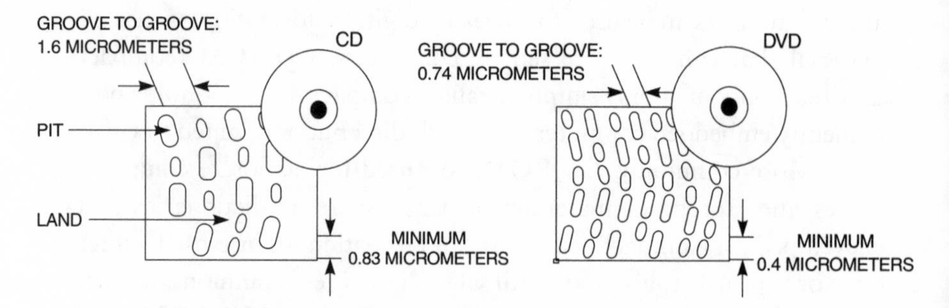

Figure 16.3 Pits and lands: compact disk and digital video disk (courtesy of Sony Electronics Inc.)

Digital Video Disks

Digital video disks have now begun to displace CDs, CD-ROMs, and videocassettes. DVD technology is essentially the same as CD technology, except that the lasers use shorter wavelengths, permitting denser storage and higher speeds. In addition, while DVD now uses only one side of each disk, the technology is defined in a way that will permit two layers of information on each side of each disk. The top layer is composed of a plastic that is transparent to the wavelength of light used to read information on the second layer (see Figure 16.4). An early single-side, single-layer 12-centimeter DVD will store 4.7 gigabytes of information, which is enough for a 90-minute motion picture, accompanying Dolby digital audio, and subtitles in thirty-two languages, all compressed using the MPEG-2 digital format, the same one used by digital satellite and terrestrial broadcasters. (A competing standard, called MO87, will start out with around 6 megabytes on each of two sides.) The DVD medium will support high-definition television, with its greater line density and wider aspect ratio than conventional TV. In its final, fully developed form, a two-layered, two-sided DVD will store 17 gigabytes, or the equivalent of four 90-minute movies. DVDs that can be recorded by users, either once or repeatedly, will also be available. The data transfer rate will be variable up to about 10 megabits per second.

A disadvantage of DVD and other (mechanical) optical media is high seek times, which generally are at least an order of magnitude longer than on hard disks. This introduces a “latency” not unlike that

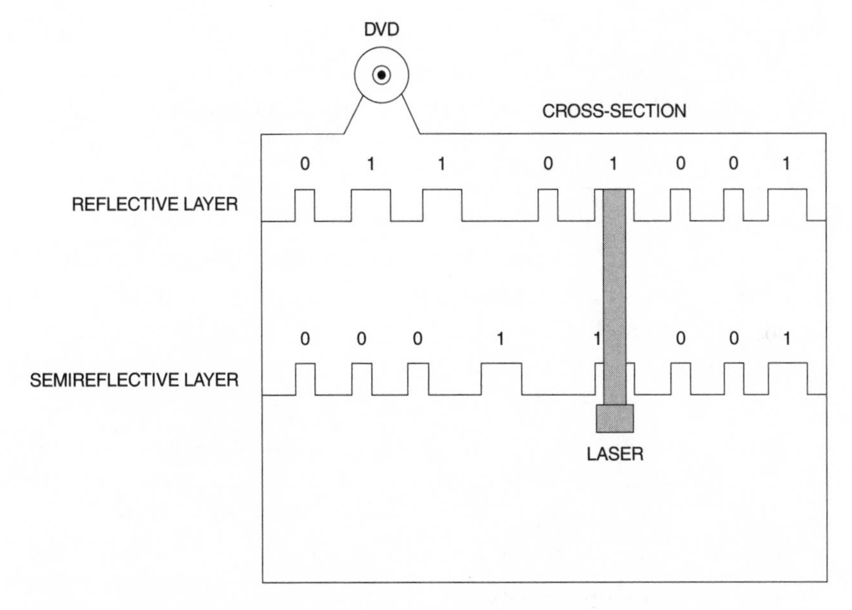

Figure 16.4 Cross-section of a digital video disk (courtesy of Sony Electronics Inc.)

found on a congested packet network or a GEO transmission. The result is that interactive uses are handicapped.

Prerecorded DVD disks will cost about $1.40 each to produce in volume, exclusive of content or distribution costs. Eight hours per day of prerecorded video will satisfy most American households if the programs are the right ones. An obvious distribution strategy is subscription DVD, in which disks are distributed by mail or in supermarkets (or with the daily newspaper). Prerecorded disks could be tailored or specialized, just like print magazines. The cost of consuming by this means the quantity of television consumed today would be perhaps $3 per day per household. This is more than current cable or DBS, which costs $1 to $2 per day per household, although with less flexible program possibilities.

Exotic Optical Storage

A key element of DVD is that it is three-dimensional: layers of data are stacked atop one another. In the trans-DVD future, there may lie even higher-density, lower-unit-cost storage devices; current research continues to focus on optical technology. One example is a project at the University of Buffalo that seeks to put 500 layers of data on a CD-ROM disk, yielding a 650-gigabyte disk with 1,000 times the capacity of a current CD-ROM. To take an even more exotic example, three-dimensional holographic storage media may be able to store orders-of-magnitude more data than can DVDs. Holographic devices store data in the form of interference patterns embedded in a three-dimensional "photorefractive" substance, such as crystal. (Holographic images can be seen on many credit cards.) If part of such a medium were destroyed, the original information would still be accessible, although its "quality" would be reduced. No other storage medium has this characteristic. Because holographic storage does not rely on bits written and read along a single line or groove, information can be stored and retrieved in two-dimensional units called pages. Such storage may be particularly useful for video on demand: each "page" might be a frame, recalled instantaneously rather than bit by bit in a scanning pattern.

As with so many of the technologies we have considered, optical storage media are decreasing in cost and increasing in capacity and capability on a scale commensurate with Moore's Law. It is not difficult to find both commercial and government projections that assume continued exponential improvement.

Holographic storage devices are not solid state. The writing and reading of data require laser beams positioned very precisely by rotating mirrors. Seek time is limited by the mechanical properties of such devices, but some researchers believe that times on the order of 10 microseconds can be achieved. Holographic technology may also be suited to the display of information in three dimensions. Early applications include displays enabling aircraft controllers to see their airspace in three dimensions, rather than a projection of three dimensions onto a two-dimensional screen. Should this technology become commercially viable, there is an obvious potential for three-dimensional entertainment displays. For example, accurate three-dimensional characters might hold forth in the middle of the living room, either in some sort of transparent block or slab, or possibly projected onto nothing but empty air. If this should happen the requirements for transmission bandwidth (and/or storage) would rapidly multiply, because representations of three-dimensional objects require far more data than an equivalent two-dimensional projection.

Beyond holographic and other optical storage technologies may lie storage in atomic lattices, where individual atoms, rather than large clusters of molecules, are the basic units. Such devices will probably be nonmechanical and extremely capacious, with very fast seek times and transfer rates.

The focus on exotic optical storage devices should not be taken as a death knell for conventional magnetic storage devices. These continue to advance, and while they approach the limits of mechanical and materials science technology at the present time, they may equal or surpass optical storage performance in the future.

As a thought experiment, imagine that technology progresses to the point that a plastic slab, a synthetic crystal the size of a deck of cards and costing $3.00 to manufacture and distribute, could hold all the information in the Library of Congress plus every movie ever produced by Hollywood. Suppose devices designed to retrieve and display information from such crystals were inexpensive and ubiquitous, and that seek times and data rates were extremely fast. What would this do to the demand for transmission bandwidth? If transmission costs remained at current levels, the effect would be to create enormous excess capacity in most digital transmission networks, including video networks. Except for real-time data flows (news and sports), everything would be shipped rather than transmitted. (Of all the programming on a 60-channel cable system, more than 90 percent is recorded months or years in advance. Most live programming is on a handful of channels—the major broadcast networks, CNN, and ESPN.) Even real-time transmissions would maximize use of caching and other local memory substitutes for bandwidth. Many communication companies would go bankrupt.

This example is whimsical in its extreme assumptions, but there is nothing fundamentally implausible about it; storage devices capable of such feats are only about two generations of technology away, and progress in storage may outpace offsetting progress in transmission and computation. Thus the key issue is not whether storage devices will reach this level of efficacy, but whether processing and transmission costs will also reach it, and when.

Conclusion

The storage of information on physical media such as tapes and disks makes it possible to use regular retail distribution channels as a substitute for electronic communication of video entertainment. The best example is a videotape rental store. If storage and retail distribution were to become cheaper or better (more diverse, higher quality) than electronic transmission media, the result might be that broadcast and other such media would be limited to real-time programming, such as news and sports.

DVD is the successor to the compact disk. In its mature form a DVD should hold eight hours or more of video programming, as much as the average household views in a day.

Exotic future storage technologies include holography, which stores information in the form of interference patterns in three-dimensional matrices. It is not impossible that because of holography and other advances, storage may win the race to be the primary consumer conduit for bulk non-real-time video programming.

CHAPTER SEVENTEEN

VIDEO ON THE WEB

It is plain to most media strategists that the Internet and video entertainment have something to do with each other. Virtually every major video network and Internet content provider has plans to integrate the two media in some fashion. Already it has become unusual for broadcast and cable networks not to offer a Web page related to their video content. Hundreds of broadcast stations also maintain Web pages. Increasingly, Web pages contain small, primitive video windows with jerky moving images transmitted in real time with poorly synchronized audio. These marginally viewable images are the result of heroic efforts to compress 20-megabyte video bitstreams into narrow 28.8-kilobit pipelines in a manner consistent with on-the-fly decompression by the user's PC.

As pointed out earlier, the Internet is ill suited as a medium for broadcasting video, and broadcasting is ill suited for providing the two-way interactive services of the Internet. Nevertheless, there are several ways to integrate the Internet and video.

Convergence? Count the Ways

First, the Internet could be used to broadcast regular TV signals, which could be displayed either on computer screens or on TV sets. With present bandwidth and compression technology, this is not possible. Still, the camel is getting smaller (due to compression) and the eye of the needle is getting bigger (due to increased capacity). Eventually, television will fit on the Internet—which doesn't necessarily mean it will end up there.

Second, Internet-transmitted low-bit-rate video information can be incorporated as a component of Web pages, for example, in a small window within the page. This is already being done by RealVideo and VDOcast.

Third, a television can be hooked up to a computer and a telephone line and used to surf the Internet, with the TV screen serving as the display. WebTV and Pippin have done this.

Fourth, the vertical blanking interval in video signals broadcast over the air or on cable can be used to transmit information to personal computers, either alone or in combination with conventional Internet access. Intel's Intercast is an example.

Fifth, conventional Web pages with content complementary to video broadcasts can be displayed (along with the full-bandwidth video) on a computer monitor or a television screen. The conventional Web pages can be transmitted either by telephone wires or in the vertical blanking interval. This is being done by Gateway 2000 and NetTV.

Sixth, information complementary to conventional broadcasts can be made available through Internet connections to servers maintained by local TV stations. Video material appears on the TV screen, Internet material appears on the computer monitor. This is the Time Warner "CityWeb" concept.

Seventh, a broadcast medium, such as cable, can be used to transmit Internet content to computer users employing one-way or two-way cable modems or the wireless equivalent. This is the @Home service, aimed at cable. Microsoft has made significant equity investments in major cable MSOs to facilitate diffusion of cable modems or set-top boxes that would utilize Microsoft software.

Eighth, narrowband nonvideo material that has the potential to satisfy the same consumer needs as conventional TV could be sent over the Internet. That is what all the jiggling, bouncing, and rotating icons on popular Web pages are about. Software such as Java and Dynamic HTML is aimed at this market.

Ninth, TV channels can be used, when they are off the air late at night, to transmit video or other information to storage devices for later access by personal computers.

Each of these alternatives, except for the first, is being experimented with at present. No one is going to send broadcast-quality video programming regularly over the Internet anytime soon. Doing so would be possible, in terms of capacity requirements, only when nonvideo Internet content begins to be delivered on whichever new digital broadband video medium survives the unfolding competition. When and if that happens, it will be a primarily video medium carrying Internet content, not the reverse.

Is such a system possible? Technically, yes. It is already being done, though in a less organized, experimental way. Indeed, it can be done in many ways. But how about costs? And consumer demand?

Assume that all national video programming is distributed by direct broadcast satellite, leaving local content to local digital services. Then if every television station in the United States had a local Web page with substantial local video content, and if the audience for such material approached the present-day audience for local TV programming (around 5 to 10 hours per TV household per week of the 50 hours of total viewing), there clearly would be insufficient capacity in local telephone systems to deliver the service. Almost any scenario in which standard-quality video is offered interactively (that is, on demand) to millions of ordinary viewers results in the collapse of present distribution systems. The interactive integrated video future requires much more capacity than we have, not only in national backbones, but in local distribution systems that link up with individual households.

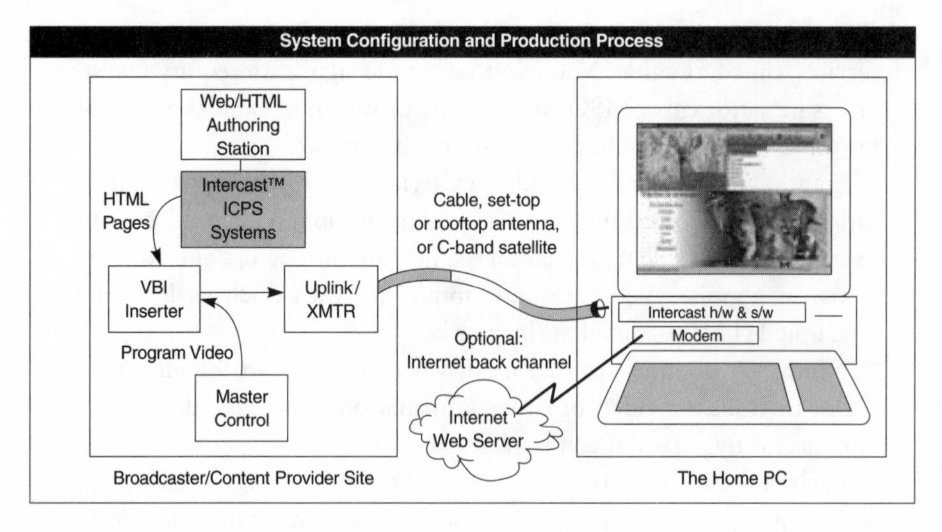

Figure 17.1 Architecture of Intercast system (Intel, 1998; used by permission)

Intercast

Intercast is an Intel technology that permits broadcasters and cable operators to use the vertical blanking interval (VBI) to send digital information to special hardware installed in personal computers (see Figure 17.1). The VBI can be used to transmit up to about 150 kilobits per second per TV channel, while leaving plenty of room for closed-captioning information and other data. The hard disk in the user's PC is used to store temporary megabytes of data that can be selectively called up by the user. The result, resembling a Web browser screen, is intended to be viewed along with the video stream on a PC, not a TV set. In effect, the Intercast data form the basis for locally cached Web pages, presumably (though not necessarily) related to the content of the video channel. User "interaction" is with the locally cached data, not with Internet sources; this is not an Internet connection. Hence there is no need for a modem or a telephone line. (The Intercast browser software, however, can be used to browse the Internet via a standard modem/telephone connection, and the VBI content may contain hyperlinks to Internet sites.) In sum, Intercast is the equivalent of the simultaneous broadcast of a video program and an ephemeral electronic magazine. WavePhore is promoting WaveTop, a device similar to Intercast. In both cases, when combined with a conventional modem back channel for upstream communications, the service is basically an asymmetric link with many of the characteristics of a cable modem or an ADSL connection. The chief difference is that the downstream capacity is relatively small, is shared by all users, and is supplemented by storage in an effort to overcome the bandwidth limitation.

@Home

A cable-television-based service called @Home Network, owned initially by a group of cable MSOs led by TCI, equipment makers, and venture capitalists and now a public company still largely controlled by TCI, offers a different concept (see Figure 17.2). @Home is a "fast home" Web page resident on servers at cable headends. That is, this is an Internet access service with a fast and powerful "value-added" set of services physically located at the cable headend, and thus extremely responsive to users. It is reached through cable modems with theoretical downstream data rates of 10 megabits per second and up. For equipment and service, users are charged about $40 per month plus installation fees. These servers offer both national and local interactive video programming. A high-speed national network, using fiber-optic cables or perhaps C-band satellite links, connects the local servers to national content sources.

The @Home service serves two related purposes. First, it is a single brand name for "cable modem" service offered by dozens of different cable companies on thousands of cable systems around the country, each of which would otherwise have to market its own brand of service. Second, it is an attempt to go beyond simple high-speed ISP access provision, permitting cable firms to offer branded content services, just like AOL and Microsoft.

As with Intercast, the @Home user who ventures beyond the local home page (the server at the cable headend) may have to contend with much slower data speeds, and the user probably will have to do without video. In a reversal of the usual state of affairs, it is the backbone rather than the last mile that constitutes the bottleneck in this system. However, unlike Intercast, @Home has the potential to add its own fiber backbone capacity and to offer a full range of ISP services.

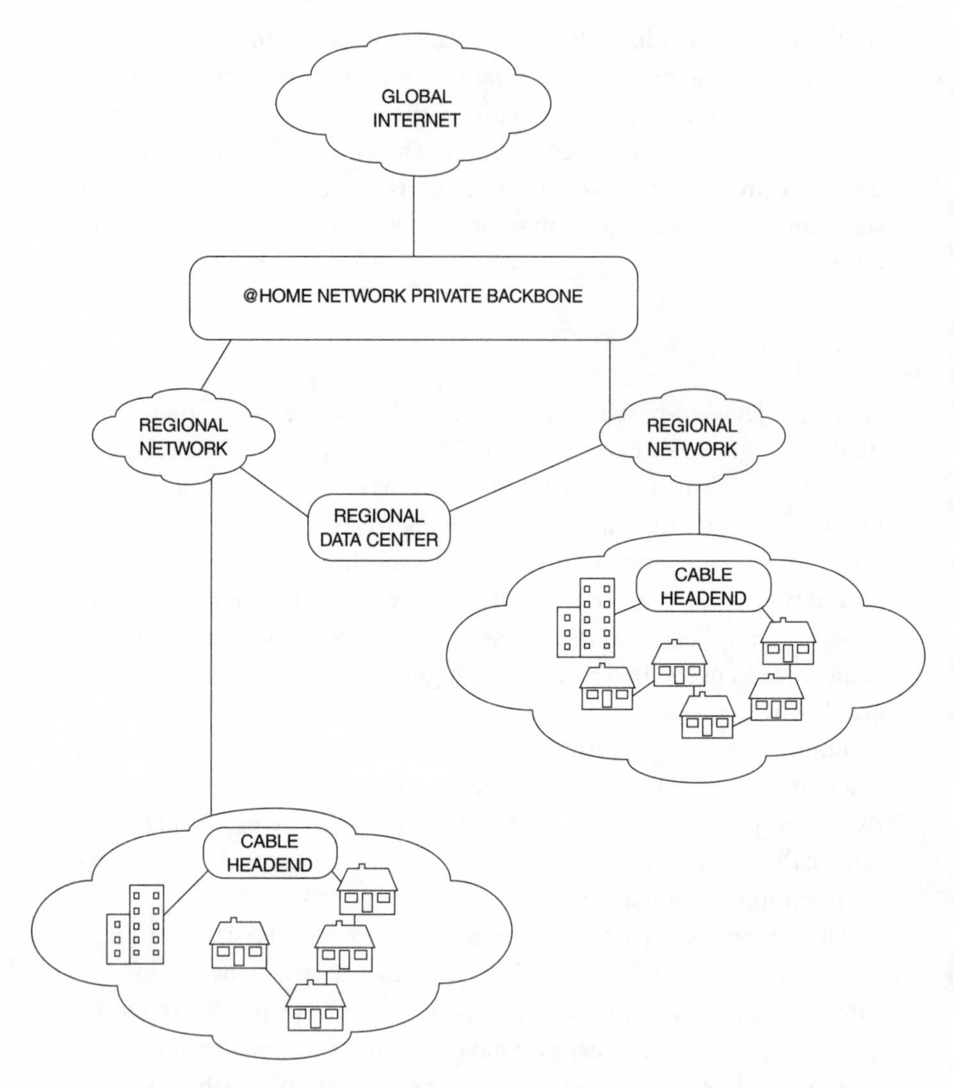

Figure 17.2 @Home network architecture

Through the use of caches and "mirroring" at the headends, speedy access to many popular Web sites can be provided. In other words, the @Home service substitutes storage nodes near the user for backbone capacity. Cox and Comcast, among other cable MSOs, have begun to offer @Home in those limited portions of their systems that have been upgraded to support cable modems. As of late 1997, however, only

25,000 subscribers had been signed up. Other MSOs, such as Time Warner and its partner U S West, offer rival services with similar features to another few tens of thousands of subscribers.

A feature of @Home worth special mention is the instant availability of content. "Connecting" to a Web page in the ordinary way means firing up one's PC, launching the browser, connecting to an Internet service provider by dialing and establishing a telephone connection, and waiting for content to begin downloading from a distant and possibly overcrowded server over a highly constrained pipeline. The process can easily take many minutes even with a fast computer and modem. In contrast, a cable modem is always "connected" (ready to share the available capacity); as a result, a number of the steps above are eliminated. If one's PC is left running continuously, the effect is almost instantaneous access to the entire Internet, just as a TV set provides nearly instantaneous access to all television channels. The cable modem puts demands on the network only when sending or receiving data, and not when in standby mode. Unlike an open telephone connection, which reserves a channel even when no data are in transit, a cable modem on standby imposes no costs on the system.

Because @Home uses cable modems and shared downstream capacity, there is a potential for overcrowding as additional users are added. The cable company can overcome this by reducing the number of subscribers connected to each node in its HFC (hybrid fiber-coaxial cable) network, but of course doing so is quite expensive. However, there is no reason in principle why an adequate quality of service cannot be sustained.

When @Home made an initial public stock offering in July 1997, it offered about 10 percent of its shares to the public. In subsequent trading, the stock price implied an overall value for the company of $1.2 billion; the company had less than $1 million in sales in 1996. @Home is aimed at the cable industry, but its technology should be applicable to any broadband local distribution system, such as LMDS. Whether cable operators will be willing to pay $300 or more per home passed for HFC upgrades in order to support one-way cable modems (more for two-way modems) depends on their assessment of how many of the passed homes will subscribe. If only 10 percent do so, for example, the cost per subscriber is a prohibitive $3,000 and up.

Multicasting

Multicasting, when used as an Internet term, means sending messages intended for multiple recipients as a single message to the last possible network node serving all recipients, and then to downstream nodes serving multiple recipients. The message is replicated as necessary at the routers. To understand this concept, suppose the Internal Revenue Service wanted to send a form letter to every local taxpayer. If all the (identical) letters were sent from Washington, tons of mail would travel by air and truck across the country. If, instead, the IRS sent one copy of the letter to the IRS office in each city, and that office printed and mailed the letter to each taxpayer, many ton-miles of transport would be saved. In the same way, multicasting saves capacity on the backbone network (as illustrated in Figure 17.3).

Multicasting technology is particularly appropriate for transmission of video bitstreams such as television programs. As a matter of fact, although multicast technology for video broadcast is relatively new on the Internet and on corporate LANs or intranets, it is basically the same as the HFC network architecture being implemented for many cable systems. Indeed, it is really the same principle that is behind C-band satellite distribution of TV network feeds to affiliates. In the case of cable, very high capacity fiber-optic cables form a backbone that connects the cable headend with neighborhood nodes. Individual households are served by coaxial cable from the nodes. Of course, today's HFC cable systems are not two-way; the nodes and households are not connected to each other, or even upstream to the headend. Multicasting on the Internet is two-way. To the extent that the Internet's two-way interactive features are employed, however, the video bitstream is no longer common to many recipients, and the benefits of multicasting are lost.

A variety of multicasting projects are already in progress. VDOcast, for example, through a joint venture with NBC and PBS, is sending live video bitstreams to suitably equipped Internet users. The joint venture focuses on video conferencing and other business-oriented services, not broadcast-quality entertainment. Mbone (Multicast backbone) is an academic multicasting experiment involving a "virtual network" that links specially programmed routers with T3 (45-megabit-per-second) pipelines. One key issue in such schemes is

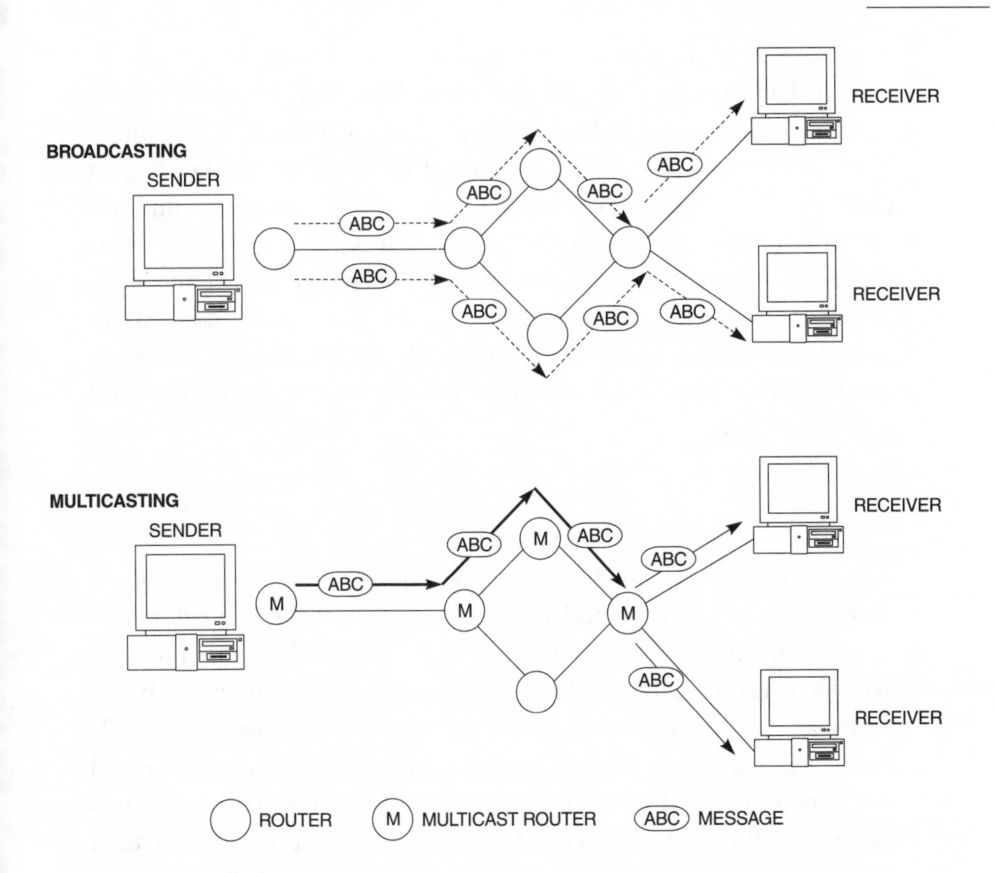

Figure 17.3 Multicasting

whether the routers, which take over some of the chores otherwise performed by servers, have the necessary capability to carry out efficiently the instructions accompanying multicast packets. Mbone has other drawbacks, including lack of compatibility with both Macintosh and PC operating systems. It runs chiefly on Unix machines. Multicasting is another example of the tradeoff between bandwidth and other factors of production in transmission, in this case processing power implemented in hardware and software at the routers.

UUnet and other ISPs now offer systems of multicast routers and servers intended to serve the commercial needs of content providers such as movie studios and TV stations. Multicasting permitted a live video webcast in conjunction with the first episode of the TV series *ER* in 1997. About 60,000 users are said to have clicked in to watch it.

Finally, as a conceptual matter, it is easy to imagine a multicast system that combines a broadcast technology with Internet technology. For example, a GEO satellite could be used to broadcast multicast content to Internet nodes at the fringes of the network, near end-users. This would save the backbone from clogging with multiple identical bitstreams, and it would alleviate latency problems. This function is of course exactly the same as that served by C-band satellites that distribute video signals to local TV stations and cable headends. In this world, the key question would be whether wired or wireless local distribution was the most efficient way to travel the "last mile" to the consumer.

WebTV and the Network Computer

The original WebTV unit had a built-in specialized computer chip, a 33.6-kilobit-per-second modem, a credit card "swiper," a keyboard, a remote control, and 2 or more megabytes of RAM. In other words, WebTV was a bare bones, limited-purpose personal computer or Internet access device hooked up to a television set. Users of WebTV used their television sets as monitors, and the company developed software to reduce the flicker and other problems associated with such use. A telephone line was required in order to reach the Internet via a WebTV-owned access service.

In 1997, after five months of operation, WebTV had fewer than 100,000 subscribers. Despite its heroic efforts to achieve ease of use, the 60 percent of the general public that does not already own a PC did not appear convinced that this or any computer is easier to deal with than the flashing clock on the VCR. So tantalizing is the prospect of integrating TVs and PCs, however, that Microsoft purchased WebTV for $425 million, and WebTV almost immediately acquired at least two direct competitors. WebTV announced late in 1997 a second-generation device that would use a 1.1-gigabyte hard disk in each home to store Web-type information downloaded late at night on TV channels for use the next day. The new model also makes a greater effort to integrate computer and TV content, at least to the extent of offering TV listings that download automatically. Still, nearly a year after Microsoft announced its purchase of WebTV, WebTV had only 250,000 subscribers.

Worldgate, a startup company backed by Citicorp and Motorola, among others, announced an effort to bring the Internet to cable viewers by means of a "dumb" set-top box with minimal local controls. Most of the brain power would be in computers located at the cable headend. Rather than rely on cable modems and expensive cable system upgrades, Worldgate would rely on the cable operator's ability to insert information in the vertical blanking interval on each channel. This approach raises the interesting property rights issue of who "owns" the right to program the VBI on broadcast stations and cable networks.

WebTV and its proliferating competitors are examples of what some see as the only hope for mass consumer penetration of the Internet: the "network computer." Household penetration of PCs appears to be holding at around 40 percent. Perhaps this is because PCs are still expensive and difficult to use. (No one wants to admit that 60 percent of all households may simply have no interest in the functions that a PC can perform, no matter how cheaply.) One proposed solution is a return to the original "dumb" remote terminal connected to a mainframe computer, high-tech fashion from the 1960s. Modern network computers would not be quite that dumb, but they would be bare bones, with as much as possible of the "intelligence" located in the network. The hope is that such computers might gain general acceptance because they might be cheap and easy to use. The association of network computers with television is simple expedience: television sets can be used as monitors, and information can be transmitted on unoccupied real estate on the channels. Further, a likely place to find the 60 percent of all households that do not use PCs is in front of their TV sets. Integration of television content and Internet content has been an afterthought—as well it might, given the disparate needs served by the two media.

Screaming Streaming Video

C-SPAN, Fox News Network, and others broadcast video on the Internet to users equipped with inexpensive, or even free, client software made by VDOnet and competing software firms. (VDO competes with similar products from RealVideo and Xing Technology, among others.) One object of the giveaway is to minimize the many disappointments likely to befall viewers of such television; more fundamentally, the idea is to promote demand for the server software and hardware that is compatible with this client software. Compression is the name of this game, along with local caching of the first few seconds of the transmission. At present the claimed frame rate is "up to" 2 or 3 per second with a 14.4 modem, and 8 to 12 per second with a 28.8 modem, compared with 30 for standard broadcast and 24 for VHS videotapes. A broadcast with VDOnet software is still subject to interruption from a slow modem, a busy backbone, or lightning storms in Arkansas.

The RealVideo technology is interesting, though it suffers from similar problems. For about $20,000 a prospective Web broadcaster can buy a RealVideo server (computer, software, and hard disk) capable of sending out 500 to 1,000 simultaneous interactive 20-kilobit-per-second video bitstreams to as many Internet surfers. The software is designed to maintain a constant number of frames per second, even in the face of Internet congestion, by "forward error correction" and "interpolation": in other words, the user's software guesses what the missing data were, based on storing a cache of frames for a few seconds on the user's hard drive before they are displayed. The standard video window is 160 by 120 pixels with a frame rate of 5 or more frames per second (a postage-stamp-size video window with a very jerky image and nonsynchronized sound). As with much Internet software, the user (client) package is very inexpensive or even free.

Yes, Virginia, There Is an Internet TV

A Gateway 2000 Internet television (which costs more than $3,000) is a combined big-screen TV and PC, equipped to surf the Web (by modem and phone line) and to display TV pictures delivered in any of the usual ways. Compaq makes something similar. These high-end devices combine a full-featured PC with a full-featured big-screen television, permitting users to display either TV programs or computer screens (or both, using multiple windows). No new function, however, is added that could not be achieved by using the two devices separately.

NetTV is a combined TV and computer that uses a computer monitor to display both. Although this is a high-end device, cheaper equivalents have long been available as plug-in modules for many different personal computers. The Macintosh computer I am using to write this has a tiny video window in the corner tuned to CNN. In 1997 Microsoft announced that the ability to receive both digital and analog video data from existing communication media would be built in to future versions of its PC operating systems, such as Windows.

Solutions

The Internet as currently engineered is not a suitable medium for the kind of television that satisfies the needs of today's viewers. The Internet lacks sufficient information-carrying capacity to do the job, and it would probably be inefficient at it even if there were enough capacity. Where can we find more capacity? Here are some of the alternatives:

- Further innovation in digital compression and some re-engineering may increase the capacity of local switched telephone systems (xDSL).
- Cable operators may settle on a cost-effective two-way digital technology, though even then an increase in capacity (changes in system engineering) would be required.
- Further innovation in digital compression and small Earth-station technology may bring affordable two-way satellite communication to individual households, just as two-way satellite service is coming for the cell phone. This probably would require changes in FCC rules regarding satellite capacity and power.
- Local television stations, employing digital transmission technologies, may offer both local and national video Internet content to local households. Like most of the current proposals to increase capacity, this service would be one-way, with a return narrowband connection using the telephone or possibly a narrowband radio link.

With almost any of these technologies, there is a tradeoff between network bandwidth and local storage capacity, and between network bandwidth and the sophistication and cost of terminal equipment. Fast local servers (perhaps themselves linked by very high speed proprietary networks) can bypass clogged national backbone networks.

Storage right at the user's PC can provide a buffer or cache that makes network delays or slow local connections more tolerable. Put a powerful enough transceiver and codec at each end, and T3 data speeds can be easily achieved. If capacity is cheap enough, even an interactive packet network (rather than a separate, one-way network) can make sense as a way to send out television broadcasts.

All of these approaches require substantial new investment by communication companies and even more substantial expenditures (by someone) on the equipment needed to receive the service in each home. Most require a significant advance in compression technology. Before anyone will undertake this investment, the potential return must be sufficient to compensate for the risks. Will the new services made possible by such investments lead to successes like radio and TV network broadcasting, personal computers, or FM stereo? Or will it lead to failed or fad services like CB radio, AM stereo, 8-track tape players, Picturephone, Betamax, or Laservision disks?

Even a successful service, such as direct broadcast satellites, may need to wait years for its moment. The basic technology for DBS has been around for almost twenty years; CBS and COMSAT announced plans, later canceled, for DBS service launches in the early 1980s. But until quite recently there wasn't an adequate program supply or a sufficiently cheap transmission and receiving system to make the service economically viable.

Conclusion

Television and the Internet can be integrated ("converge") in a variety of ways. The Internet could carry conventional TV signals, Web pages can incorporate small video windows, television sets can be used as computer monitors, broadcasters can insert Internet data in the vertical blanking interval, Web pages can carry content complementary to conventional TV channels, and cable modems can connect households with the Internet, among other possibilities. Some of these, especially the first, require investments to increase capacity, but none is technically infeasible. In addition, the Web may compete with television simply by offering both consumers and advertisers another nonvideo alternative.

Several new services seek to combine television and the Internet.

@Home is a cable industry effort to offer both access service and Web content to cable subscribers. Multicasting and other technologies are being developed that reduce the burden on the Internet backbone of transmitting television broadcasts. Intel offers a system designed to make use of the VBI. WebTV is an ambitious effort to solve the user acceptance problems associated with the Internet. The consumer purchases a bare-bones computer hooked to her telephone and television. The device permits use of the TV screen as a monitor for Web surfing and other functions. But after almost a year on the market, WebTV had sold less than 100,000 units.

In effect, the current offerings are either low-fixed-cost speculative ventures or mere experiments. As yet, no successful integration of television and the Internet seems in the offing. The more promising approaches all require massive, risky investments in fixed-cost infrastructure.

CONCLUSION

The analog communication world is in decline, overtaken in important respects by digital technology. We abandon analog for digital media for no aesthetic reason, but in simple pursuit of economy. Digital media are more economically efficient than analog media, and so we use them more. But analog media are not inherently less efficient; indeed, the reverse is true because analog media do not require the extra steps of digitizing and then decoding. Next week or next year a lurch in science and engineering could produce an analog communication technology superior to digital technology. It happened that digital processors developed more rapidly than analog ones. The key was cost.

Communication costs fell and therefore prices did as well. When prices fall, society consumes more of the cheaper service. We are going to be consuming more communication, both broadcast and narrowcast, and at least for the immediate future this communication will take digital forms. That does not necessarily mean we will do all our communicating through a single pipeline. Costs and prices are falling because of technological progress in processing and transmission and because of increased supplies of spectrum from the government, not merely from economies of scale in sharing pipelines.

Although communication channels in the immediate future will all be digital, network architectures will vary. No one architecture can cheaply satisfy all the demands of consumers for different functions. Specifically, there is no reason to suppose that a network optimized to carry non-real-time narrowband messages (for example, e-mail) will lend itself to real-time broadband transmissions. The Internet is not going to carry television broadcasts, at least not in anything like their current form. The Internet has the wrong network structure, and Internet pipelines are too small for television. The Internet, however, may eventually carry competing entertainment services that reduce the television audience. These services are more likely to be successful if they are more passive than interactive. It is passive, not interactive, entertainment for which consumers have displayed the strongest demand.

Paul Saffo of the Institute for the Future, Neal Stephenson, a prize-winning science fiction writer, and others foresee the rise of "nanotechnology," an integration of analog and digital forms (Saffo, 1997a,b; Stephenson, 1995). They believe sensors and other analog microelectromechanical systems (MEMS), integrated with digital electronics on chips, are of growing importance. These devices sense, for example, temperature, lighting, and other conditions in order to trigger environmental controls in buildings. Such a sensor triggers the operation of automobile airbags. Currently nanotechnology has no special relevance for communication. But it is interesting to speculate that in ten years we are just as likely to be in the age of nanotechnology as in the sixth or seventh decade of the digital age, as we now perceive it.

It Would Be Magical

A mime performs on a street corner, her crimson hat on the sidewalk in hope of coins. Merlin appears; a word of magic power is spoken and the mime is projected onto a thousand street corners, performing above a thousand hats. This is television.

Merlin then peers into his crystal ball and calls up images of the past and future. He surveys all that is known and exchanges wisdom with other wizards. This is the Internet.

In many ways it does not matter how Merlin works his magic. What is really interesting are the limits of magic. How many mimes, at what cost, can Merlin project? How many mimes will the government permit Merlin to project? How much faster can the crystal ball gain access to information? How much does Merlin have to pay for his crystal ball, and to whom?

It would be astonishing—magical—if it turned out that analog broadcasting, digital broadcasting, GEO direct broadcast satellites, Teledesic, wireless cable, conventional cable, telephone lines souped up with digital technology, and digital video disks were equally effective and profitable ways to transmit the Cooking Channel. Most likely, one will be more cost effective than the others. Similarly, it would be astonishing if all of these media were equally effective and profitable ways for consumers to interact with airline reservation systems. No doubt, one will be better than the others. Finally, there is no reason why the media that are best at these two tasks should be the same, although they might share certain common resources.

Media convergence makes sense as a hypothesis if there are strong economies of scope in having a number of functions served by a single pipeline, or if the use of a single pipeline does not very much reduce the value of the services that are provided, compared with their value when provided separately. To forecast that television and the Internet will converge is to make speculative assumptions about these factors.

Costs: Who Has the Advantage?

It is futile to attempt to analyze in detail the costs and capabilities of the rival video media of the future. Many of the costs are subject to drastic unpredictable change, and the capabilities are no more permanent. It is for this reason that numerous investors are holding off commitments to particular technologies. Cable companies are only tiptoeing into cable modems, telephone companies are backing off video service, and broadcasters are resisting the FCC's urgings to go digital.

Consider, for a moment, what would happen if all the elements of communication were free (had zero costs). Would everything go down one big digital pipe? Surely not, for network topologies have features other than cost that affect demand. There would be as many different networks as there are combinations of network characteristics that matter to consumers. If communication were free, what communications would take place? Surely video entertainment on demand would be important, but so would prescheduled programming that could be viewed passively, such as current network offerings. Interactive Internet services would be more popular than they are today because of lower cost and much less delay, but without changes in content and interface, Internet use would probably not be a mainstream activity. Thus even if one assumes that communication is free, it is by no means easy to predict the shape of mass media consumption. That costs are most definitely not zero makes the problem even more complicated. Nevertheless, a quick summary of the important costs and capabilities is enlightening.

The cost characteristics of the rival media divide into two groups: network costs and terminal equipment costs. Terminal equipment is the new box that sits on the TV or next to the computer, and perhaps the antenna that goes with it. Network costs are those related to building and launching satellites, installing fiber-optic cables, or constructing radio transmitters. All of the scenarios are about the same in one respect: they involve expensive new terminal equipment. Network cost characteristics, however, vary greatly. A geosynchronous direct broadcast satellite—or even a LEO system—has huge up-front costs but virtually no variable costs until 100 percent capacity utilization is reached. The same is true for the Internet backbone. Such networks have substantial common costs, shared by many users, thus reducing the cost per user. At the other extreme is the telephone system with its star-shaped networks. Expanding the capacity of any subscriber line is just like adding terminal equipment, and the costs quickly become astronomical.

The most promising digital media and their current advantages, disadvantages, and estimated costs are summarized in the Appendix. In order to enhance the comparability of the data, and despite the loss of realism, all costs are stated as "greenfield" costs—that is, the cost of constructing each system from scratch rather than as an incremental improvement to an existing system. Further, costs are stated on the basis of 100 percent coverage of U.S. television households. Because the costs look very different if penetration levels less than 100 percent are achieved, both the fixed cost (per household, given 100 million households served) and the variable cost per household are provided.

Even with these adjustments, comparisons remain difficult because the systems have different attributes. Some systems (for example, conventional GEOs) cannot yet provide local content. Others (for example, digital video disks) cannot provide live content. Some (certain digital subscriber lines) are fully two-way interactive video systems that would permit, say, video conferencing and the operation of home video servers in connection with personal Web pages. Others (such as NVOD or asymmetric VOD) permit only limited interactivity.

Which system is cheapest? The answer depends on which set of services or capabilities is the minimum required to attract consumers. And this is no easier to determine. Two observations, however, can be made. First, digital television as currently proposed does not seem to be a very attractive system compared with several others. Second, terrestrial wireless systems such as local multichannel distribution service (LMDS) are extremely appealing, chiefly because of their low fixed costs. These wireless systems do not require massive up-front investments, which lowers the financial risks considerably.

All of the video delivery systems have one thing in common: they cost more than the average American household is now willing to pay for video services, even video services integrated with Internet content.

The Appendix also includes network configurations, because network topology is no less important a component of video delivery systems than bandwidth, processing, and storage. Star wireless broadcast networks have by far the cheapest unit costs for one-way video delivery. Among such networks there is a tradeoff between the very low per-unit fixed costs of GEO satellites and the very high per-unit fixed costs of hundreds of terrestrial broadcast stations, each offering local content. Among terrestrial broadcast technologies, multichannel topologies clearly dominate single-channel (or fewer-channel) multistation topologies. If two-way video service is an important feature, then switched broadband networks are required. They do not currently exist, and are hideously expensive to build. Narrowband switched networks such as the present telephone system or the Internet are not fully capable of handling one-way, much less two-way, switched video.

High-Stakes Poker

Which if any of these scenarios will come to pass? They all cost about the same, and they all offer features, and suffer handicaps, that are very difficult to evaluate because of the absence of information about consumer demand. Analysts with the latest engineering and marketing information can get out their spreadsheet programs and do the calculations, but the range of uncertainty about the assumptions overwhelms any attempt to conclude that one of these scenarios is superior to (more profitable than) the others, especially in light of changing technology and the effects on costs of network externalities. For example, a relatively small improvement in compression technology could make narrowband telephone (the present system converted to DSL) the long-term medium of choice for both Internet and even conventional video.

No supplier wants to be in the middle of a multibillion-dollar investment project, only to have a small change in the technology make its efforts obsolete. So both the risks and the stakes are enormous.

This explains, no doubt, the difficulties that the big cable television MSOs have been having with the new technologies. In 1994 and 1995, some cable operators announced plans for advanced digital interactive service to all of their subscribers (so-called 500 channel systems). Since then, notwithstanding several experimental projects, the plans have been repeatedly delayed or canceled. Two forces are at work. First, cable is in a fight for its life with direct satellite broadcasting, an effort that puts pressure to bear on cable cash flows due to increased promotional spending and program upgrades. Cable MSOs must fight it out with DBS or abandon the field by turning cable into a cash cow with a finite life. Second, no particular technology looks like the clear winner in maintaining cable's current advantage as a purveyor of local content, while permitting it to offer the relevant range of new interactive and conventional video services.

Over-the-air broadcasters face a problem that is no less difficult than the challenges before cable operators. They must settle on a business plan for exploiting the new digital airwave assignments, and they must do so heavily constrained by the political baggage that goes with continued regulation. For example, the broadcasters may be required to maintain some form of free over-the-air television service. In addition, they will be restricted in their ability to combine their airwave assignments into local units of efficient scale, in order to offer a service with capacity comparable to that of DBS, cable, LMDS, or 38 gigahertz. As with cable, any new digital broadcast service will require substantial investments in consumer equipment. Traditionally, consumers must be persuaded to make this expenditure. So rapid is the pace of change that broadcasters (and other providers) may offer the equipment at heavily subsidized prices to induce consumers to subscribe. To do so, of course, raises the financial stakes enormously. There are nearly 100 million TV households, and if each one were to be supplied with just one cable modem at $500 per unit, the cost would be $50 billion. If every household were to be offered just one xDSL telephone line at $1,000 per line, the cost would be $100 billion.

These sums require established firms to "bet the company" on a particular technology. In some ways entrepreneurs and startups with access to venture capital may be in a better position to make these gambles than less nimble and more risk-averse communication giants. No matter how the tale unfolds, it will be a high-stakes poker game. Every player wants every possible edge—not just the "right" technology, but government protection and first-mover advantages. The coming years will be every bit as exciting as the early days of radio. It is even possible that in some lonely laboratory an apostle of Philo T. Farnsworth, TV's inventor, has found a way to fit the camel that is television through the eye of the Internet needle.

APPENDIX

Medium	Comment	Status	Suitable for two-way interactive video? Suitable for near video on demand (NVOD)?	Est. cost to serve 100 million U.S. household subscribers, each		Capacity (TV channels available to each household with 5:1 digital compression)	Local content?
				Fixed cost of system	Variable cost per household		
Terrestrial							
Terrestrial wireless							
Conventional analog VHF and UHF[a]	Today's broadcasters	Supposedly will disappear in 5–10 years	No/No	$160	Zero	Average is 12	Yes
Digital television (DTV)[b]	Each broadcaster gets one channel	To be phased in over 5–10 years	No/No	$160	$200–$3,000	Average is 60	Yes
Analog multichannel multipoint distribution systems (MMDS)	Weak showing so far		No/No	$50	$500	33 max.	Yes
Digital MMDS	Telesis Los Angeles system operational	Auctions planned	No/Yes	$50	$500–$700	160+	Yes

Digital LMDS	Propagation issues, e.g. weather problems	Auctions planned	Perhaps/Yes	$50	$500–$700	830 channels	Yes
Digital 38 gigahertz	Propagation issues; frequency reuse	Auctions planned	Yes/Yes	$50	Unknown	Up to 2,500 without frequency reuse	Yes
Terrestrial wired							
Copper pairs (telephone)[c]							
Analog modems (to 56 kilobits per second)	How most people gain access to the Internet; not suited for video	56 kilobits per second is probably max. speed	Yes/Yes (but extremely slow and low quality)	0	<$200	Up to 1 percent of one channel	Yes
Digital modems							
Integrated services digital network (ISDN)	Speeds up to 128 kilobits per second	Widely available	Yes/Yes (but extremely slow and very low quality)	$500–$3,000	$400	2 percent of one channel	Yes

Medium	Comment	Status	Suitable for two-way interactive video? Suitable for near video on demand (NVOD)?	Est. cost to serve 100 million U.S. household subscribers, each		Capacity (TV channels available to each household with 5:1 digital compression)	Local content?
				Fixed cost of system	Variable cost per household		
Digital subscriber lines (xDSL)	Speeds up to T1 (1.554 megabits per second)	Experimental	Yes/Yes (but slow and low quality)	$1,000+	$600	Less than one channel (VHS-quality video)	Yes
Cable television							
Conventional	Passes 90 percent of U.S. households, two-thirds subscribe		No/Yes	0	Negligible	Up to 120	Yes
Digital upgrade (cable modems)	Technology uncertain	Several experiments	Yes/Yes	$500–$1,000+	$500	Up to 500	Yes
Space Systems							
Geosynchronous (GEO)							
Conventional (C-band)[d]	Requires very large antennas	Obsolete	Not practical for residential use	$15	$2,000	Thousands	No
Digital (Ku-band)[e]	DSS, Primestar, Echo Star	Growing rapidly	Yes/Yes (but two-way very expensive)	$75	$200+	Thousands	No

Digital with local stations	Murdoch 1997 proposal	May never be built	Yes/Yes (but two-way very expensive)	$100+	$200+	Thousands	Yes
Low Earth orbit (LEO) satellites							
Iridium (Motorola)	Designed for mobile phones	Several launches have taken place	No/No (insufficient bandwidth)	$6 billion	$500	1 or 2 percent of one channel	Not a video medium
Teledesic (McCaw/Gates)	Designed for Internet connections up to T1	To be launched in 2001	Yes/Yes (but slow and low quality)	$9 billion to serve hundreds of thousands of simultaneous users	Unknown	Less than one channel (VHS quality video)	Possible but unlikely in practice
Storage							
CD-ROM	Real-time transmission not possible	Exists	No/Yes	0	$3,000 per household per year	Unlimited	Yes
DVD[f]	Real-time transmission not possible	Introduced in 1997	No/Yes	0	$730 per household per year	Unlimited	Yes

Medium	Comment	Status	Suitable for two-way interactive video? Suitable for near video on demand (NVOD)?
Network configurations			
Point to points (broadcast)			No/Yes
Switched channel (telephone)			Yes/Yes (if enough bandwidth)
Packet switched (Internet)			No/No (latency problems)

a. Programming and conventional TV set costs ignored. Assumes 1,600 stations at construction cost of $10 million each.

b. Assumes 1,600 stations at construction cost of $10 million each. HDTV sets cost up to $3,000. Set-top boxes will cost at least $200.

c. Telephone-based delivery offers the opportunity for video on demand (VOD), so that, while only one channel may be transmitted at a time, the user can select the content of that channel from whatever menu of options is maintained by the sender. LEO systems operate similarly.

d. Assumes 25 satellites at $60 million each, plus $2,000 each for antenna and equipment.

e. Assumes 10 satellites at $75 million each.

f. DVD (which once stood for "digital video disk" or "digital versatile disk" and now stands for nothing) cost to supply 8 hours of video per day per household by distributing disks, assuming distribution cost of $2 per disk.

GLOSSARY

ADSL	Asymmetrical digital subscriber line. The most popular technology for transferring large amounts of data over ordinary, but short, local phone wires. Accomplished by installing special equipment on the ends of the lines. The volume of data flowing downstream to the user is much greater than upstream flow. *See* DSL.
Analog data	Data represented by a physical quantity that is continuously variable and proportional to the data.
ARPAnet	Predecessor to the Internet. Developed by Defense Department in 1969.
ATM	Asynchronous transfer mode. An advanced data transmission and switching protocol that greatly increases the capacity of transmission paths, both wired and wireless. ATM uses packets of fixed size and establishes "virtual" circuit connections.
ATV	Advanced television. An FCC term to designate what is now called DTV or digital television. *See* HDTV.
Backbone	That portion of a communication network, such as the Internet, made up of very high capacity trunks connecting switches or routers.
Bandwidth	The difference, in hertz, between the highest and

	lowest frequencies of a transmission channel. (Also used loosely—but not herein—to refer to the capacity of a communication medium.)
Bit	A binary unit of information or data derived from a choice between two equally probable alternatives, such as zero or one, on or off.
Bitstream	A sequence of bits transmitted on a communication channel.
Broadband	(1) A high-capacity communication link, wired or wireless, capable of transmitting the equivalent of multiple TV signals. *See* Narrowband. (2) Any communication channel or medium capable of data rates in excess of what can be achieved with a telephone line and an analog modem.
Broadcast	One-to-many communications, print or electronic. *See* Multicast; Narrowcast.
Broadcasting	As used herein, radio or television (video) transmissions.
Browser	Software that facilitates examination and retrieval of information in distant databases on the World Wide Web.
Buffer	A mechanism for storing data temporarily because they are arriving faster than they can be processed.
Byte	A defined number of bits, usually eight, often corresponding to a letter or symbol, upon which computer operations are performed.
Cable modem	A device that permits one-way or two-way high-speed data communication over a cable television system for purposes such as Internet access.
Cache	A temporary store of data intended for use or reuse; for example, recently viewed Web pages that might be revisited.
C-band	A portion of the electromagnetic spectrum designated by the FCC for, among other things, the first commercial satellite communications.
CD	A (digital) compact disk, originally for music; also used for computer data, in which case it is called a CD-ROM, for "read-only memory."
Cellular system	A wireless communication system in which relatively low-power or focused transmitters reuse frequencies in noncontiguous geographic areas (cells).
Circuit switching	A communication network in which users are con-

	nected, through switches, using a channel dedicated to that use for the duration of the communication. A telephone system is an example of such a network. Packet networks, in contrast, are "connectionless."
Clarke orbit	The orbit at an altitude of 22,300 miles above the equator at which a satellite is stationary relative to the Earth.
Client	A computer or user in communication with a server.
Coaxial cable	A broadband transmission line consisting of two cylindrical copper conductors arranged concentrically, separated by insulation.
Codec	Coder-decoder or compressor-decompressor. Hardware or software that serves as an intermediary between a computer and a digital transmission medium.
Compression	Reduction of the bandwidth or number of bits needed to encode information, most commonly by eliminating redundant bits. A means of saving transmission time and storage space.
Convergence	The hypothesis that a single digital medium such as the Internet will replace analog telephone and television media.
Data	A collection of bits. The quantities or symbols on which computer and communication equipment operate, typically stored or transmitted in the form of electromagnetic energy. *See* Bit; Information.
DBS	Direct broadcast satellite television service, such as DirecTV or Primestar. A.k.a. direct to home (DTH). *See* GEO.
Digital	A function that operates in discrete steps, such as "on" and "off." Because the physical world is continuous, such representations are approximations. Digital communications uses discontinuous, discrete electrical, optical, or electromagnetic signals that change in frequency, polarity, or amplitude.
DirecTV	A DBS service operated by General Motors' Hughes business unit. Shares a satellite with USSB; the combined system is called DSS.
Download	The process of retrieving data from a distant database; also, the data so retrieved.
DRAM	A type of random access memory chip.

DSL — Digital subscriber line. An ordinary telephone line improved by expensive equipment, making it capable of broadband transmissions. DSL comes in many flavors, known collectively as xDSL. *See* ADSL, VDSL.

DTV — Digital television. The term adopted by the FCC to describe its specification for the next generation of broadcast television transmissions. DTV encompasses both HDTV and STV. *See* ATV, HDTV, STV.

DVD — Originally, digital video disk or digital versatile disk; now stands for nothing. Physically similar to a CD, a DVD is much more densely packed with data. Eventually it will contain the equivalent of eight hours of TV programming.

e-mail — Text messages created and viewed on PCs and transmitted electronically, usually over an office network or over the Internet.

Ethernet — A protocol for transmitting computer data over local area networks.

Fiber-optic cable — A cable containing one or more optical fiber strands. Each strand is capable in theory of carrying 25 trillion bits per second.

Fiber optics — Hair-thin glass strands allowing laser beams to be bent and reflected with low levels of loss and interference. A.k.a. "glass optical wave guides" or "optical fibers."

FTP — File transfer protocol. A procedure for transmitting files of computer data over the Internet.

GEO — Geosynchronous communication satellite in the Clarke orbit at an altitude of 22,300 miles. It remains in a fixed position relative to the Earth.

Gigahertz — One billion hertz (q.v.). 28 and 38 gigahertz: portions of the spectrum designated by the FCC for terrestrial broadband fixed services. 28 gigahertz is currently used for LMDS; 38 gigahertz for wireless trunks.

HDTV — High-definition television. A digital TV transmission with the same letterbox shape and about the same clarity as a theatrical motion picture. *See* DTV.

Headend — The control center of a cable television system, where incoming signals are amplified, converted, processed, and combined into a common coaxial or optical cable for transmission to subscribers.

hertz	The frequency in cycles per second of a wave form or carrier used for communication.
HFC	Hybrid fiber-coaxial system. A local cable TV or telephone distribution network consisting of fiber-optic trunks ending at neighborhood nodes, with coaxial cable feeder and drop lines downstream of the nodes.
HTML	HyperText Markup Language. Defining characteristic of the World Wide Web.
HTTP	HyperText Transfer Protocol. Standard for transferring documents on the World Wide Web.
Hypermedia	A nonlinear representation of information that allows users to access related works or images from a single computer screen. For example, a user reading an encyclopedia entry on jazz music can also hear excerpts from recordings and view photos of various artists. Sometimes synonymous with "multimedia."
Hypertext/hyperlink	Text, symbols, or icons which, when actuated by the user, establish a connection to a database, which may be local or on the World Wide Web.
Information	In communication theory, a measure of one's freedom of choice in selecting a message, or of the range of possible alternatives when receiving a message. The greater the number of potential messages, the greater the information contained in any one. Not to be confused with data or meaning.
Interlace scan	A technique for capacity conservation in TV broadcasting. A conventional television frame, composed of 480 active lines, is transmitted as two sequential fields: one containing the odd lines and the other the even lines. Because there are 60 fields per second, this creates some of the visual effect of higher frame rates.
Internet	The physical connections through which millions of computer users exchange data. The Internet comprises thousands of smaller networks, each associated with an organization such as a firm, a university, a government agency, or an ISP. Communication is possible because of voluntary agreements to use certain communication techniques. *See* WWW.

Intranet	Interconnected IP networks confined within an organization, enterprise, or membership group; Intranets may be connected to the Internet.
IP	Internet protocol. The standards describing data packets, addressing, and router behavior that make interconnection among networks possible.
ISDN	Integrated Services Digital Network. An early and limited version of a digital subscriber line with capacity of either 64 or 128 kilobits per second. ISDN is viewed by some as a technological bridge between the current telephone system and an updated, broadband network. Others see ISDN as a symbol of the failure of local telephone companies to adapt promptly to new technology.
ISO	International Standards Organization. A voluntary organization that coordinates industrial standards organizations in dozens of member countries. In the United States, the FCC mandates compliance with certain voluntary ISO standards.
ISP	Internet service provider. An organization that arranges connections between the Internet and individuals or enterprises. Large ISPs operate their own Internet backbones. Value-added ISPs (AOL, CompuServe) offer information services as well as interconnection.
Ka-band	A portion of the electromagnetic spectrum reserved by the FCC for both terrestrial and satellite uses, for which various specific proposals have been made.
Ku-band	A portion of the electromagnetic spectrum designated by the FCC for, among other things, direct broadcast satellites.
LAN	Local area network. Communication paths linking computers, printers, and servers into a network for use by an individual, office, school, or other organization.
Latency	Delay. The interval between transmission and receipt of a message or between command and execution of an instruction. Latency arises from mechanical delay, such as drive-seek time, from limited processor speeds, from the finite speed of light, and from network capacity constraints.
LEO	Low Earth orbit satellite systems. Communication satellites in orbit at altitudes of a few hundred miles. Each system has dozens or even hundreds of

satellites, each of which is in constant motion relative to any point on the Earth. *See* GEO.

LMDS — Local multichannel distribution service, a.k.a. "wireless cable." A new broadband wireless service operating in a frequency range (28 gigahertz) designated by the FCC for that purpose.

Local loop — The pair of dedicated copper wires (or equivalent channel) that connects each telephone to a local switch or "central office." *See* Twisted pair.

MEMS — Microelectromechanical systems, a.k.a. "smart matter." MEMS combine microprocessors with tiny devices such as sensors, valves, gears, mirrors, and actuators embedded in semiconductor chips. Some believe MEMS nanotechnology, rather than multimedia, will be the key technology of the early twenty-first century.

MMDS — Multichannel multipoint distribution system, a.k.a. "wireless cable." A local wireless terrestrial video broadcast technology that relies on line-of-sight transmission. An analog MMDS has up to 33 television channels; a digital MMDS may have 150 or more.

Modem — Modulator/demodulator. A device that transforms digital information into analog form for transmission over analog telephone lines, reversing the process for received data.

Moore's Law — The quantity of microelectronic processing speed, power, or memory that can be purchased with a dollar doubles every two years or so. In contrast, Internet traffic doubles two to five times per year. Named for Gordon Moore of Intel.

MPEG — Motion Picture Experts Group. An ISO-related industry standards organization that develops standards for coding video transmissions.

MSO — Multiple system operator. Any company that owns a large number of cable television systems.

Multicast — (1) A procedure for minimizing the Internet backbone capacity requirements of broadcasting identical simultaneous bitstreams to multiple recipients. No more than one stream is sent to any node serving two or more recipients.
(2) The use of digital spectrum assignments by broadcasters to air multiple channels of "standard" television rather than one channel of HDTV.

Multimedia	A term with so many and varied usages that it is nearly meaningless. Most commonly, a computer equipped with a CD-ROM drive and speakers.
Must-carry rule	A statutory requirement that cable television operators must purchase the rights to use local TV signals or, at the option of the TV station, must carry the signal compulsorily.
Narrowband	A low-capacity communications link, such as a telephone cable, which with present technology is incapable of transmitting multiple TV signals. *See* Broadband.
Narrowcast	Anyone-to-anyone switched communications.
Network	The collection of links that connects end-users with one another and with devices such as servers, switches, and routers.
Network effect	The economic effect that arises when the value of a good or service to any user increases as the number of other users increases. Also known as "network externality." Fax machines are a good example.
Network topology	The arrangement of links (and switches) in a communication network. Networks can be hierarchical, star-shaped, looped, connectionless, switched, broadcast, and so on.
NTSC	National Television Standards Committee. Used to refer to the technical standards and physical characteristics of conventional analog TV broadcasting, as enshrined in FCC regulations.
NVOD	Near video on demand. The practice of offering a given movie to subscribers on multiple channels. Starting times are staggered, such as every thirty minutes.
Orbital slot	An arc segment of the Clarke orbit assigned by the FCC to a particular licensee. Slots are divided among countries by international agreement. Like spectrum, U.S. orbital slots are now auctioned.
Packet switching	A "connectionless" network protocol. Data are assembled into packets of variable size (for example, 200 bytes), which include both the source and the destination addresses. Packets are transmitted to routers, which contain instructions from which efficient routing decisions can be computed. Packets in the same message may travel by different routes

	to their common destination. Packets are reassembled by the recipient's computer.
PC	Personal computer.
PCS	Personal communications services. An FCC term for digital cell phones using recently auctioned frequencies.
Polarization	The direction (such as horizontal or vertical) of the electrical field in an electromagnetic wave. Signals with different polarizations can be sent simultaneously on the same frequency.
POTS	Plain old telephone service.
Primestar	A direct broadcast satellite service owned by a consortium of major cable television operators.
Protocol	A formal description of the message formats and rules that computers, switches, or other devices must follow when exchanging messages.
Public good	A good or service, such as a TV program, whose quantity or value is not reduced by consumption.
RAM	Random-access computer memory.
Router	A network device that determines the optimal paths along which network traffic should be forwarded. Routers forward packets to other routers or to other networks. Occasionally called a "gateway."
RSVP	Resource reservation protocol. A method—involving hardware and software—that improves transmission of real-time bitstreams on the Internet.
Search engine	Software that facilitates the discovery of relevant information in distant databases.
Server	A network-connected computer (and associated storage device, such as a hard disk) that contains data intended to be accessed by distant users.
Set-top box	An electronic device that mediates between any, but especially a digital, television distribution system (cable, satellite, wireless) and an ordinary analog television set.
Shannon's Law	The quantity of accurate information that can flow over a channel has an upper limit determined by the bandwidth of the channel and its signal-to-noise ratio. Named for Claude Shannon of Bell Labs.
SMATV	Satellite master antenna television. A private cable television system serving an apartment complex or

similar residential grouping. Such systems currently serve around one million United States households.

SONET — Synchronous optical network. A protocol used in intercity trunks composed of fiber cables.

Spectrum — The range of wavelengths (or frequencies) of electromagnetic radiation, from the longest radio waves to the shortest gamma rays. Visible light is only a small part of this range. A.k.a. "frequency spectrum" or "airwaves."

Streaming — A real-time bitstream conveying audio or video information.

STV — Television of "standard" as opposed to "high-definition" quality under the FCC's 1997 digital television standards.

Switch — A device that can establish temporary physical connections between a large number of pairs of communication channels.

T1, T3, etc. — Telephone industry designations for broadband trunks or connections of various capacities. (OC is used to designate fiber-optic media.)

Transponder — One of the (twelve or more) active electronic units in a communication satellite. Receives signals from Earth, translates to a different frequency, amplifies, and then broadcasts (downlinks) them.

Trunk — A physical or wireless broadband connection linking switches or routers to one another. Each trunk carries many connections. In contrast, a local loop generally connects a single subscriber with a nearby central office switch.

Twisted pair — Ordinary telephone wires used to transmit between the telephone instrument and the local office. *See* Local loop.

UHF — Ultra high frequency. A range of spectrum designated by the FCC for television broadcasts; originally channels 14 to 84, later reduced to 14 to 69.

VBI — Vertical blanking interval. A portion of the signal used to broadcast conventional TV programs that is not used for video information and hence is available to transmit other data, such as captions.

vBNS — Very high speed backbone network service. The next generation of Internet facilities sponsored by NSF, also known as Internet II.

VCR	Videocassette recorder.
VHF	Very high frequency. The original FCC-designated television broadcast spectrum. Channels 2 to 13.
VOD	Video on demand. A service permitting the user to view a movie or other video program, selected from a large menu of such choices, at any time.
Web page	An element of the interface offered to the user by a distant database, as displayed on the user's computer monitor when running a Web Browser program.
Wireless cable	*See* Gigahertz; LMDS; MMDS. Wireless cable systems of all kinds served fewer than one million United States households in 1997.
WWW	World Wide Web, or Web. That very popular and growing portion of the Internet devoted to data transfers mediated by hyperlinks and Web browser software. ("Web" was formerly a slang term for a broadcast network such as ABC.)
xDSL	*See* DSL; ADSL.

REFERENCES

Adams, William J., and J. L. Yellen. 1976. "Commodity Bundling and the Burden of Monopoly." *Quarterly Journal of Economics* 90: 475–498.

Alpert, Mark. 1992. "CD-ROM: The Next PC Revolution." *Fortune* 125 (13): 68–73.

Anderson, Robert H., et al. 1995. *Universal Access to E-mail: Feasibility and Societal Implications*. Santa Monica, Calif.: Rand Corporation.

Arthur, W. B. 1983. "On Competing Technologies and Historical Small Events: The Dynamics of Choice under Increasing Returns." Working Paper. Laxenburg, Austria: Institute for Applied Systems Analysis.

Auerbach, Jon G. 1997. "Getting the Message." *Wall Street Journal,* June 16, p. R22.

Auletta, Ken. 1991. *Three Blind Mice: How the TV Networks Lost Their Way.* New York: Random House.

Austin, Bruce A. 1990. "Home Video: The Second-Run 'Theater' of the 1990s." In *Hollywood in the Age of Television,* ed. Tino Balio. Boston: Unwin Hyman.

Aztech Company. 1997. "Aztech Launches High Performance 10x Speed CD-ROM Drives."
URL: http://www.aztech.com.sg/whatnew/10xcdrom.htm.

Baran, Paul. 1962. *On Distributed Communications Networks.* Santa Monica, Calif.: RAND Corporation.

Baran, Paul, Sharla P. Boehm, and Joseph W. Smith. 1964. *On Distributed Communications.* Vols. I–XI. *Memorandum.* Santa Monica, Calif.: RAND Corporation.

Barnow, E. A. 1966–1970. *History of Broadcasting in the United States.* New York: Oxford University Press.

Barwise, Patrick, and Andrew Ehrenberg. 1988. *Television and Its Audience.* London: Sage Publications.

Besen, S. M. 1987a. *New Technologies and Intellectual Property: An Economic Analysis.* Note N-2601-NSF. Santa Monica, Calif.: RAND Corporation. May.

———. 1987b. "Some New Standards for the Economics of Standardization in the Information Age." In *Economic Policy and Technological Performance,* ed. P. Dasgupta and P. L. Stoneman. New York: Cambridge University Press.

Besen, S. M., et al. 1984. *Misregulation Television: Network Dominance and the FCC.* Chicago: University of Chicago Press.

Besen, S. M., and Robert Crandall. 1981. "The Deregulation of Cable Television." *Law and Contemporary Problems* 44: 79–124.

Besen, S. M., and L. L. Johnson. 1986. *Compatibility Standards, Competition, and Innovation in the Broadcasting Industry.* Prepared for the National Science Foundation R-3453-NSF. Santa Monica, Calif.: RAND Corporation.

Besen, S. M., and G. Saloner. 1989. "The Economics of Telecommunications Standards." In *Changing the Rules: Technological Change, International Competition, and Regulation in Communications,* ed. R. Crandall and K. Flamm. Washington, D.C.: Brookings Institution.

Besen, S. M., and R. Soligo. 1973. "The Economics of the Network-Affiliate Relationship in the Television Broadcasting Industry." *American Economic Review* 63: 259–268.

Bilby, Kenneth. 1986. *The General.* New York: Harper & Row.

Borcherding, T. E. 1978. "Competition, Exclusion, and the Optimal Supply of Public Goods." *Journal of Law and Economics* 21: 111–132.

Breyer, Stephen. 1982. *Regulation and Its Reform.* Cambridge, Mass.: Harvard University Press.

Brinkley, Joel. 1997. *Defining Vision: The Battle for the Future of Television.* New York: Harcourt Brace and Co.

Bush, Vannevar. 1945. "As We May Think." Atlantic Monthly 176: 101–108.

C-Cube Microsystems. 1996. "Compression Technology: An MPEG Overview."
URL: http://www.c-cube.com/technology/mpeg.html#MPEG Overview.

Carter, T. B., M. A. Franklin, and J. B. Wright. 1986. *The First Amendment and the Fifth Estate.* Mineola, N.Y.: Foundation Press.

Caruso, Denise. 1992. "New Consumer Devices Are Coming." *Digital Media.* July 22.

Cerf, Vinton G. 1989. "Requiem for the ARPANET." In *Users' Dictionary of Computer Networks,* ed. Tracy LaQuey. Bedford, Mass.: Digital Press.

Cheshire, Stuart. 1996. "The Quest for Interactivity." (Web site no longer exists.)

Ciciora, Walter S. 1995. "Cable Television in the United States: An Overview." Boulder, Colo.: CableLabs.

Clift, C., and A. Greer. 1989. *Broadcast Programming*. Lanham, Md.: University Press of America.

Clinton, William J., and Albert Gore, Jr. 1997. *A Framework for Global Electronic Commerce*. Released by the White House, July 1.

Coase, Ronald H. 1950. *British Broadcasting: A Study in Monopoly*. Cambridge, Mass.: Harvard University Press.

———. 1959. "The Federal Communications Commission." *Journal of Law and Economics* 2: 1–40.

———. 1962. "The Interdepartmental Radio Advisory Committee." *Journal of Law and Economics* 5: 17–47.

———. 1972. "Durability and Monopoly." *Journal of Law and Economics*. 15: 143–149.

Corbato, Fernando J., et al. 1992. "The Project MAC Interviews." *In IEEE Annals of the History of Computing*, 14(2), ed. John A. N. Lee and Robert Rosin.

Cottrell, Lee. 1997. Internet End-to-End Performance Monitoring, Methodology, Tools, and Results. Stanford Linear Accelerator Center. URL: http://www.slac.stanford.edu/grp/scs/net/talk/xiwt-sep97/sldoo2.htm.

Crandall, Robert W., and Harold Furchtgott-Roth. 1996. *Cable TV: Regulation or Competition?* Washington, D.C.: Brookings Institution.

Daniel, Stephen, James Ellis, and Tom Truscott. 1980. *USENET—A General Access UNIX Network*. Durham, N.C.: Duke University Press.

Daniels, George H. 1970. "The Big Questions in the History of American Technology." *Technology and Culture* 11 (January): 1–21.

David, P. A. 1975. *Technical Choice, Innovation, and Economic Growth: Essays on American and British Experience in the Nineteenth Century*. Cambridge: Cambridge University Press.

———. 1986. "Understanding the Economics of QWERTY: The Necessity of History." In *Economic History and the Modern Economist*, ed. W. N. Parker. Cambridge, Mass.: Basil Blackwell.

de Sola Pool, Ithiel, ed. 1977. *The Social Impact of the Telephone*. Cambridge, Mass.: MIT Press.

———. 1983. *Technologies of Freedom*. Cambridge, Mass.: Harvard University Press.

———. 1990. *Technology without Boundaries: On Telecommunications in a Global Age*. Cambridge, Mass.: Harvard University Press.

De Vany, A. S., et al. 1969. "A Property System for Market Allocation of the Electromagnetic Spectrum: A Legal-Economic-Engineering Study." *Stanford Law Review* 21: 1499–1561.

Demsetz, H. 1970. "The Private Production of Public Goods." *Journal of Law and Economics* 13: 293–306.

Denning, Peter J., and Robert M. Metcalfe. 1997. *Beyond Calculation: The Next Fifty Years of Computing.* New York: Springer-Verlag.

Dertouzos, Michael. 1997. *What Will Be: How the New World of Information Will Change Our Lives.* San Francisco: HarperEdge.

Dunnett, P. 1990. *The World Television Industry.* New York: Routledge.

Dutton, W. H. 1995. "Driving into the Future of Communications?—Check the Rear View Mirror." In *Information Superhighways: Multimedia Users and Futures,* ed. S. J. Emmott. London: Academic Press.

Dutton, W. H., and M. Peltu, eds. 1997. *Information and Communication Technologies: Visions and Realities.* New York: Oxford University Press.

Easystor Company Web Site. 1997.
URL: http://www.easystor.com/html/fdd120.html.

Egan, Bruce L. 1996. *Information Superhighways Revisited.* Boston: Artech House.

Ellis, David. 1996. *Split Screen: Home Entertainment and the New Technologies.* Friends of Canadian Broadcasting. URL: http//:friendscb.org.

Fano, Robert. 1961. *Transmission of Information.* New York: MIT Press and John Wiley & Sons.

Farrell, J., and G. Saloner. 1985. "Standardization, Compatibility, and Innovation." *RAND Journal of Economics* 16: 70–83.

Faulhaber, G. R. 1987. *Telecommunications in Turmoil: Technology and Public Policy.* Cambridge, Mass.: Ballinger Publishing.

Find/SVP. 1995. "Internet Use Impedes Other Leisure Activities." December.

———. 1997. The 1997 American Internet User Survey. URL: http://etrg.findsvp.com.

Fischer, Claude S. 1992. *America Calling: A Social History of the Telephone to 1940.* Berkeley: University of California Press.

Fisher, F. 1985. "The Financial Interest and Syndication Rules in Network Television: Regulatory Fantasy and Reality." In *Antitrust and Regulation: Essays in Memory of John J. McGowan,* ed. F. Fisher. Cambridge, Mass.: MIT Press.

Fogg, Chad. 1996. Questions That Should Be Frequently Asked about MPEG. URL:
http://bmrc.berkeley.edu/projects/mpeg/faq/mpeg2-v38/ faq_v38.html#tag1.

Fournier, G. M. 1986. "The Determinants of Economic Rents in Television Broadcasting." *Antitrust Bulletin* 31: 1045–1066.

Fowler, M. S., and D. L. Brenner. 1982. "A Marketplace Approach to Broadcast Regulation." *Texas Law Review* 60: 207–257.

Freeman, W. H. 1966. *Information.* San Francisco: Scientific American.

Georgia Tech University. 1993–1996. "The American Internet User." GVU Internet Survey, versions 1–7.

Gigi Information Services, BIS Consumer Electronics Information Service. 1995. "Colour Television Formats and Features—USA." Pp. 16–37.

Gilder, George. 1989. *Microcosm*. New York: Touchstone.

———. 1991. "Now or Never." *Forbes* 148(8): 188–198.

Gladwell, Malcolm. 1997. "Dept. of Disputation." *New Yorker,* April 7, 1997.

Glick, I., and S. Levy. 1962. *Living with Television*. Chicago: Aldine.

Goldenson, L. H. 1991. *Beating the Odds*. New York: Charles Scribner's Sons.

Hafner, Katie, and Matthew Lyon. 1996. *Where Wizards Stay Up Late: The Origins of the Internet*. New York: Simon & Schuster.

Hartmanis, Jaris, and Herbert Lin. 1992. *Computing the Future: A Broader Agenda for Computer Science and Engineering*. Washington, D.C.: National Academy Press.

Hatfield Associates, Inc. 1997. "Hatfield Model Release 3.1." Prepared for AT&T and MCI.

Hauben, Michael. 1994/1995. "The Vision of Interactive Computing and the Future." *Amateur Computerist* 6 (2/3): 3–6.

Hauben, Ronda. 1994. "From ARPANET to Usenet News." *Amateur Computerist* 5 (3/4): 1–10; 6 (1): 14–16; and 6 (2/3): 19–20.

Hazlett, Thomas W. 1989. "Cabling America: Economic Forces in a Political World." In *Freedom in Broadcasting,* ed. C. Veljanovski. London: Institute of Economics.

———. 1990a. "Duopolistic Competition in Cable Television." (And "Reply.") *Yale Journal on Regulation* 7(65): 119, 141–148.

———. 1990b. "The Rationality of U.S. Regulation of the Broadcast Spectrum." *Journal of Law and Economics* 33:133–175.

———. 1991. "The Demand to Regulate Franchise Monopoly: Evidence from CATV Rate Deregulation in California." *Economic Inquiry* 29(2): 275–296.

———. 1997a. "Physical Scarcity, Rent-Seeking, and the First Amendment." *Columbia Law Review* 97(4): 905–944.

———. 1997b. "Prices and Outputs under Cable TV Reregulation." *Journal of Regulatory Economics* 12(2): 173–193.

Hazlett, Thomas W., and M. L. Spitzer. 1997. *Public Policy towards Cable Television,* vol. 1: *The Economics of Rate Controls*. Cambridge, Mass.: MIT Press.

Head, S., and C. H. Sterling. 1990. *Broadcasting in America: A Survey of Electronic Media,* 6th ed. Boston: Houghton Mifflin Co.

Heart, Frank, et al. 1978. *The ARPANET Completion Report*. Washington, D.C.: BBN.

Herring, J. M., and G. C. Gross. 1936. *Telecommunications: Economics and Regulation*. New York: McGraw-Hill.

Hilmes, Michele. 1990. *Hollywood and Broadcasting: From Radio to Cable.* Chicago: University of Illinois Press.

Hoskins, C., and R. Mirus. 1988. "Reasons for the U.S. Dominance of the International Trade in Television Programmes." *Media, Culture, and Society* 10: 499–515.

In Memoriam: J.C.R. Licklider, 1915–1990. 1991. Palo Alto, Calif.: Digital Systems Research Center, Digital Equipment Corp.

Inglis, Andrew F. 1990. *Behind the Tube: A History of Broadcasting Technology and Business.* Boston: Focal Press.

Innis, H. 1951. *The Bias of Communication.* Toronto: University of Toronto Press.

Institute for Information Studies. 1997. *The Internet as Paradigm.* Queenstown, Md.: Nortel and Aspen Institute.

Intel. 1998. Intercast Technology Hybrid Application Cookbook. URL:
http://developer.intel.com/drg/hybrid_author/cookbooks/intercast/03.htm.

Internet Performance Working Team. 1997. Notes.
URL: http://www.xiwt.org/xiwt/documents/.

Internet Society News. Various issues. Reston, Va.: Internet Society.

Johnson, L. L. 1970a. *Cable Television and the Question of Protecting Local Broadcasting.* Report R-595-MF. Santa Monica, Calif.: RAND Corp.

———. 1970b. *The Future of Cable Television: Some Problems of Federal Regulation.* Memorandum RM-6199-FF. Santa Monica, Calif.: RAND Corporation.

Johnson, L. L., and D. R. Castleman. 1991. *Direct Broadcast Satellites: A Competitive Alternative to Cable Television?* Santa Monica, Calif.: RAND Corporation.

Johnson, L. L., and D. Reed. 1990. *Residential Broadband Services by Telephone Companies? Technology, Economics, and Public Policy.* Santa Monica, Calif.: RAND Corporation.

Johnson, N. 1970. *How to Talk Back to Your Television Set.* New York: Bantam Books.

Johnson, Steve. 1997. *Interface Culture: How New Technology Transforms the Way We Create and Communicate.* San Francisco: HarperEdge.

Joskow, P. L., and N. L. Rose. 1989. "The Effects of Economic Regulation." In *The Handbook of Industrial Organization,* ed. Richard Schmalensee and Robert D. Willig. Amsterdam: North Holland Press.

Kalakota, Ravi, and Andrew B. Whinston. 1996. *Frontiers of Electronic Commerce.* Reading, Mass.: Addison-Wesley.

Katz, M. L., and C. Shapiro. 1986. "Technology Adoption in the Presence of Network Externalities." *Journal of Political Economy* 94: 822–841.

Keirstead, Phillip, O. Keirstead, and Sonia-Kay Keirstead. 1990. *The World of Telecommunication: Introduction to Broadcasting, Cable, and New Technologies.* Boston: Focal Press.

Kelly, Kevin. 1994. *Out of Control: The New Biology of Machines, Social Systems, and the Economic World.* Reading, Mass.: Addison-Wesley.

Keyes, J. 1997. *Webcasting: How to Broadcast to Your Customers over the Net.* New York: McGraw-Hill.

Keynote Systems. 1997. Keynote Byusiness 40. URL: http://www.keynote.com.

Kleinrock, Leonard. 1976. "On Communications and Networks." *IEEE Transactions on Computers* C-25(12): 1320–1329.

Krasnow, E. G., and L. D. Longley. 1973. *The Politics of Broadcast Regulation.* New York: St. Martin's Press.

———. 1978. *The Politics of Broadcast Regulation.* 2d ed. New York: St. Martin's Press.

Krasnow, E. G., L. D. Longley, and H. A. Terry. 1982. *The Politics of Broadcast Regulation.* 3d ed. New York: St. Martin's Press.

Krattenmaker, Thomas G. 1997. *Telecommunications Law and Policy.* 2d ed. Durham, N.C.: Carolina Academic Press.

Krattenmaker, Thomas G., and Lucas A. Powe, Jr. 1994. *Regulating Broadcast Programming.* Washington, D.C.: American Enterprise Institute.

Kubey, Robert, and Mihaly Csikszentmihalyi. 1990. *Television and the Quality of Life: How Viewing Shapes Everyday Experience.* Hillsdale, N.J.: Lawrence Erlbaum Associates.

Landler, Mark. 1997. "The Logic of Losing at All-News TV." *New York Times,* June 22, p. E4.

Lang, K. 1957. "Areas of Radio Preferences: A Preliminary Inquiry." *Journal of Applied Psychology* 41: 7–14.

Larson, Erik. 1992. "Watching Americans Watch TV." *Atlantic* 269(3): 66–80.

Levin, H. J. 1971. *The Invisible Resource: Use and Regulation of the Radio Spectrum.* Baltimore, Md.: Johns Hopkins University Press.

Levy, Mark. 1989. *The VCR Age: Home Video and Mass Communication.* Newbury Park, Calif.: Sage Publications.

Lewis, Tom. 1991. *Empire of the Air: The Men Who Made Radio.* New York: HarperCollins.

Licklider, J. C. R. 1960. "Man-Computer Symbiosis." *IRE Transactions on Human Factors in Electronics* HFE-1: 4–11.

———. 1965. *Libraries of the Future.* Cambridge, Mass.: MIT Press.

Licklider, J. C. R., and Robert Taylor. 1968. "The Computer as a Communication Device." *Science and Technology: For the Technical Men in Management* 76: 21–31.

Licklider, J. C. R., and Albert Vezza. 1978. "Applications of Information Networks." *Proceedings of the IEEE* 66 (11): 43–59.

Lim, Joe S. 1998. "Digital Television: Here at Last." *Scientific American.* May. 78–83.

Malamud, Carl. 1992. *Exploring the Internet: A Technical Travelogue.* Englewood Cliffs, N.J.: Prentice Hall.

Manishin, G. 1987. "Antitrust and Regulation in Cable Television: Federal Policy at War with Itself." *Cardozo Arts and Entertainment Law Journal* 6: 75–100.

———. 1990. "An Antitrust Paradox for the 1990s: Revisiting the Role of the First Amendment in Cable Television." *Cardozo Arts & Entertainment Law Journal* 9:1–14.

Marill, Thomas, and Lawrence G. Roberts. 1966. "Toward a Cooperative Network of Time-Shared Computers." *Proceedings, Fall Joint Computer Conference* 29: 425–431.

McCarthy, John. 1966. "Information." In *Information, A Scientific American Book*. San Francisco: W. H. Freeman.

———. 1992. "John McCarthy's 1959 Memorandum." *IEEE Annals of the History of Computing* 14(1): 20–23.

McChesney, Robert W. 1993. *Telecommunications, Mass Media, and Democracy.* New York: Oxford University Press.

McGowan, J. J. 1967. "Competition, Regulation and Performance in Television Broadcasting." *Washington University Law Quarterly* 44: 499–520.

McKnight, L. W., and J. Bailey, eds. 1997. *Internet Economics.* Cambridge, Mass.: MIT Press.

McLuhan, Marshall. 1964. *Understanding Media: The Extensions of Man.* New York: McGraw-Hill.

McLuhan, Marshall, and Q. Fiore. 1996. *The Medium Is the Message: An Inventory of Effects.* San Francisco: Hardwired.

Merrill Lynch. 1996. In "From Couch Potato to Cybersurfer." *Economist* 340 (7973): 72.

MIDS. 1997. Internet Weather Report.

Minasian, J. R. 1969. "The Political Economy of Broadcasting in the 1920s." *Journal of Law and Economics* 12: 391–403.

Minoli, Daniel. 1997. *Internet and Intranet Engineering: Technologies, Protocols, and Applications.* New York: McGraw-Hill.

Minow, Newton. 1961. "The Vast Wasteland." Speech delivered to the National Association of Broadcasters, May 9, 1961. Reprinted in N. Minow, *Equal Time: The Private Broadcaster and the Public Interest.* New York: Atheneum, Macmillan Publishing, 1964.

Mirabito, Michael M., and Barbara L. Morgenstern. 1990. *The New Communications Technologies.* Boston: Focal Press.

Motorola Corporation. 1997. "Application for Authority to Construct, Launch, and Operate the Celestri Multimedia LEOSystem [as filed with the FCC]."
URL: http://www.mot.com/GSS/SSTG/projects/celestri/ASD_product.html.

Nadel, Mark S. 1983. "COMCAR: A Marketplace Cable Television Franchise Structure." *Harvard Journal on Legislation* 20: 541–573.

National Cable Television Association. 1997. "Directory of Cable Television Networks." *Cable Television Developments.* Washington, D.C.: NCTA, p. 6.

National Opinion Research Center. 1977–1994. "General Social Survey." Combined Annual Surveys.

National Research Council. 1996. *The Unpredictable Certainty: Information Infrastructure through 2000.* Washington, D.C.: National Academy Press.

National Science Foundation. Office of the Inspector General. 1993. *Review of NSFNET.* Washington, D.C.

National Science Foundation. Office of the Inspector General. 1996. *Future of NSFNET.* Washington, D.C. (Web site no longer exists.)

NCA Computers Corporation.
URL: http://www.ncacomputers.com/2e.hdexternal.html#scsicdroms.

Negroponte, N. 1995. *Being Digital: A Roadmap for Survival on the Information Superhighway.* London: Hodder & Stoughton.

Nelson, Ralph, and Sidney Winter. 1982. *An Evolutionary Theory of Economic Change.* Cambridge, Mass.: Harvard University Press.

NetSpeed, Inc. 1997. "General Introduction to Copper Access Technologies." URL: http://www.netspeed.com/tutorial.html.

Neuman, W. R. 1992. "The Technological Convergence: Television Networks and Telephone Networks." In *Television for the 21st Century: The Next Wave.* Aspen Institute Program on Communications and Society.

Noam, Eli M. 1987. "A Public and Private-Choice Model of Broadcasting." *Public Choice* 55: 163–187.

———. 1991. *Television in Europe.* New York: Oxford University Press.

Noll, A. Michael. 1997. *Highway of Dreams: A Critical View along the Information Superhighway.* Mahway, N.J.: Lawrence Ehrlbaum Associates.

Noll, Roger G. 1989. "The Economic Theory of Regulation after a Decade of Deregulation, Comments and Discussion." In *Brookings Papers on Economics Activity, Microeconomics 1989,* ed. S. Peltzman, M. N. Baily, and C. Winston. Washington, D.C.: Brookings Institution.

Noll, Roger G., and B. M. Owen. 1983. *The Political Economy of Deregulation: Interest Groups in the Regulatory Process.* Washington, D.C.: American Enterprise Institute.

——— 1988. "United States v. AT&T: The Economic Issues." In *The Antitrust Revolution,* ed. J. E. Kwoka Jr. and L. J. White. Glenview, Ill.: Scott Foresman.

———. 1989. "United States v. AT&T: An Interim Assessment." In *Future*

Competition in Telecommunications, ed. S. P. Bradley and J. A. Hausman. Boston: Harvard Business School Press.

Noll, Roger G., M. J. Peck, and J. J. McGowan. 1973. *Economic Aspects of Television Regulation.* Washington, D.C.: The Brookings Institution.

Office of Science and Technology Policy. 1989. *The Federal High Performance Computing Program.* Washington, D.C.

Owen, Bruce M. 1970. "Public Policy and Emerging Technology in the Media." *Public Policy* 18: 539–552.

———. 1975. *Economics and Freedom of Expression: Media Structure and the First Amendment.* Cambridge, Mass.: Ballinger Publishing Co.

———. 1978. "The Economic View of Programming." *Journal of Communication* 28: 43–47.

Owen, Bruce M., and Ronald Braeutigam. 1978. *The Regulation Game: Strategic Use of the Administrative Process.* Cambridge, Mass.: Ballinger Publishing Co.

Owen, Bruce M., and Peter R. Greenhalgh. 1986. "Competitive Policy Considerations in Cable Television Franchising." *Contemporary Policy Issues* 4: 69–79.

Owen, Bruce M., and Steven Wildman. 1992. *Video Economics.* Cambridge, Mass.: Harvard University Press.

Pacey, P. L. 1985. "Cable Television in a Less Regulated Market." *Journal of Industrial Economics* 34: 81–91.

Park, R. E. 1972. "Cable Television, UHF Broadcasting, and FCC Regulatory Policy." *Journal of Law and Economics* 15: 207–231.

———. 1973. "New Television Networks." Report R-1408-MF, Santa Monica, Calif.: RAND Corporation.

———. 1975. "New Television Networks." *Bell Journal of Economics* 6: 607–620.

Peltzman, Sam. 1989. "The Economic Theory of Regulation after a Decade of Deregulation." In *Brookings Papers on Economic Activity, Microeconomics 1989,* ed. S. Peltzman, M. N. Baily, and C. Whinston. Washington, D.C.: Brookings Institution.

Pierce, John R. 1972. "Communication." *Scientific American* 227(3): 31–41.

Poltrack, David. 1983. *Television Marketing.* New York: McGraw-Hill.

Posner, Richard A. 1971. "Taxation by Regulation." *Bell Journal of Economics and Management Science* 2: 22–50.

Powe, L. A., Jr. 1987. *American Broadcasting and the First Amendment.* Berkeley: University of California Press.

Roberts, Lawrence G. 1978. "The Evolution of Packet Switching." *Proceedings of the IEEE* 66(11): 1307–1313.

———. 1988. "The ARPANET and Computer Networks." In *A History of Personal Workstations,* ed. A. Goldberg. New York: ACM Press.

Robinson, Glen O. 1978. "The Federal Communications Commission: An

Essay on Regulatory Watchdogs." *University of Virginia Law Review* 64: 169–262.

Robinson, John P., and Geoffrey Godbey. 1997. *Time for Life: The Surprising Ways Americans Use Their Time.* University Park, Pa.: Pennsylvania State Press.

Rosse, James N. 1967. "Daily Newspapers, Monopolistic Competition, and Economies of Scale." *American Economic Review Papers and Proceedings* 57: 522–533.

Rothenberg, J. 1962. "Consumer Sovereignty and the Economics of TV Programming." *Studies in Public Communication* 4: 45–54.

Saffo, Paul. 1997a. "Sensors: The Next Wave of Innovation." *Communications* of the ACM.
URL: http://www.acm.org/pubs/periodicals/cacm/FEB97/ saffo.html.

———. 1997b. "Sensors: The Next Wave." *Cadence Archive.* URL: http://www.cadence.com/features/archive/vol2no1/saffo.html.

———. N.d. "Sensors: The Next Wave of Infotech Innovation." Menlo Park, Calif.: Institute for the Future.
URL: http://www.itftf.org/sensors/sensors.html.

Samuelson, Paul A. 1954. "The Pure Theory of Public Expenditure." *Review of Economics and Statistics* 36: 387–389.

———. 1955. "Diagrammatic Exposition of a Theory of Public Expenditure." *Review of Economics and Statistics* 37: 350–356.

———. 1958. "Aspects of Public Expenditure Theories." *Review of Economics and Statistics* 40: 332–338.

Schmookler, Jacob. 1962. "Economic Sources of Inventive Activity." *Journal of Economic History* 22(1):1–20.

Schroeder, Erica. 1996. "Pinnacle Micro's Apex Rivals Speed, Cost of Hard Drives." *PC Week.*
URL: http://www.pcweek.com/archive/31/pcwk0012.htm.

Schumpeter, Joseph. 1950. *Capitalism, Socialism, and Democracy.* 3d ed. New York: Harper Bros.

Setzer, Florence, and Jonathan Levy. 1991. *Broadcast Television in a Multichannel Marketplace.* OPP Working Paper Series 26. Washington, D.C.: Federal Communications Commission. June.

Shannon, Claude E. 1948. "The Mathematical Theory of Communication." *Bell System Technical Journal* 27: 379–423, 623–656.

Shannon, Claude E., and W. Weaver. 1964. *The Mathematical Theory of Communication.* Urbana: University of Illinois Press.

Shenk, David. 1998. *Data Smog: Surviving the Information Glut.* San Francisco: Harper.

Smiley, A. K. 1990. "Regulation and Competition in Cable Television." *Yale Journal on Regulation* 7: 121–139.

Smith, Laura B. 1996. "Storage Central." *PC Week.*
URL: http://www.pcweek.com/archive/960520/pcwk0078.htm.

Smith, Ralph Lee. 1972. *The Wired Nation*. New York: Harper & Row.

Smith, Sally B. 1990. *In All His Glory: The Life of William S. Paley*. New York: Simon & Schuster.

Spence, A. Michael. 1976. "Product Selection, Fixed Costs and Monopolistic Competition." *Review of Economic Studies* 43: 217–235.

Spence, A. Michael, and Bruce M. Owen. 1977. "Television Programming, Monopolistic Competition, and Welfare." *Quarterly Journal of Economics* 91: 103–126.

Spitzer, Mark L. 1991. "The Constitutionality of Licensing Broadcasters." *New York University Law Review* 64: 990–1071.

Stefik, Mark. 1996. *Internet Dreams: Archetypes, Myths, and Metaphors*. Cambridge, Mass.: MIT Press.

Steiner, Peter O. 1952. "Program Patterns and Preferences, and the Workability of Competition in Radio Broadcasting." *Quarterly Journal of Economics* 66: 194–223.

Stephenson, Neal. 1995. *The Diamond Age*. New York: Bantam.

Technology Futures, Inc. 1996. Company Web Site.

Tele.com. Various issues. URL: http://www.telecomdot.com.

Teledesic Corporation. 1996. "The Importance of Latency in Satellite Networks."

Thompson, Earl A. 1968. "The Perfectly Competitive Production of Collective Goods." *Review of Economics and Statistics* 50: 1–12.

Toshiba Corporation. 1997. "Toshiba DVD Players Lead the Way in Performance, Features, and Flexibility."
URL: http://www.toshiba.com/tacp/PressRelease/dvd22.html.

Turkle, Sherry. 1996. *Life on the Screen: Identity in the Age of the Internet*. New York: Simon & Schuster.

TV & Cable Factbook. Annual. Washington, D.C.: Warren Publishing.

U.S. Department of Commerce. Bureau of the Census. 1975. *Historical Statistics of the United States, Colonial Times to 1970*. Washington, D.C.: Bureau of the Census, p. 798.

U.S. Department of Commerce. Bureau of the Census. Annual, 1981–1996. *Statistical Abstract of the United States*.

U.S. Department of Commerce. Bureau of the Census. 1997. "Resident Population of the United States: Estimates by Age and Sex." Consistent with 1990 Census.
(www.census.gov/population/estimates/nation/intfile2–1.txt)

U.S. Department of Commerce. National Technical Information Service. 1996. "JTEC Panel Report on Optoelectronics in Japan and the United States." Report PB96–152202.

U.S. Department of Commerce. National Telecommunications and Information Administration. 1988. *Video Program Distribution and Cable Television: Current Policy Issues and Recommendations*.

———. National Telecommunications and Information Administration.

1995. *Survey of Rural Information Infrastructure Technologies.* NTIA Special Publication 95–33.

U.S. Federal Communications Commission. Annual, 1994–. *Annual Assessment of the Status of Competition in the Market for the Delivery of Video Programming.*

U.S. Federal Communications Commission. 1980. Report of the Network Inquiry Special Staff. *New Television Networks: Entry, Jurisdiction, Ownership, and Regulation.*

U.S. Federal Communications Commission–Government Accounting Office, "Survey of U.S. Cable Systems." In Pike and Fischer, Inc., 1990. *Cable Television Service (Competition and Rate Deregulation Policies)* 43, no. 38: 1831–1843.

U.S. Office of Telecommunications Policy. 1974. *Report to the President by the Cabinet Committee on Cable Communications.*

U.S. President's Task Force on Communications Policy. 1968. *Final Report.*

Varian, Hal R. 1995. "Pricing Information Goods." Paper presented at Research Libraries Group Symposium on Scholarship in the New Information Environment. Harvard Law School.

Varian, Hal R., and Richard Roehl. 1996. "Circulating Libraries and Video Rental Stores."

Vogel, Harold L. 1986. *Entertainment Industry Economics: A Guide for Financial Analysis.* Cambridge: Cambridge University Press.

———. 1990. *Entertainment Industry Economics: A Guide for Financial Analysis.* 2d ed. New York: Cambridge University Press.

Vogelsang, Ingo, and Bridger M. Mitchell. 1997. *Telecommunications Competition: The Last Ten Miles.* Washington, D.C.: American Enterprise Institute.

Waterman, David. 1985. "Prerecorded Home Video and the Distribution of Theatrical Feature Films." In *Video Media Competition,* ed. E. M. Noam. New York: Columbia University Press.

Webster, J. G. 1986. "The Television Audience: Audience Behavior in the New Media Environment." *Journal of Communication* 36: 77–91.

Webster, J. G., and J. J. Wakshlag. 1983. "A Theory of Television Program Choices." *Communication Research* 10: 430–446.

Werbach, Kevin. 1997. *Digital Tornado: The Internet and Telecommunications Policy.* FCC Office of Plans and Policy Working Paper 29. Washington, D.C.

Wiener, Norbert. 1948. *Cybernetics: Or Control and Communication in the Animal and the Machine.* New York: John Wiley and Sons.

———. 1994. *The Legacy of a Centennial Symposium.* Cambridge, Mass.: MIT Press.

Wildman, Steven S., and B. M. Owen. 1985. "Program Competition, Diversity, and Multichannel Bundling in the New Video Industry." In *Video*

Media Competition: Regulation, Economics, and Technology, ed. E. M. Noam. New York: Columbia University Press.

Wiles, P. 1963. "Pilkington and the Theory of Value." *Economic Journal* 73: 183–200.

Wilson, Kevin G. 1988. *Technologies of Control: The New Interactive Media for the Home.* Madison: University of Wisconsin Press.

Wood, James. 1996. *Satellite Communications Pocket Book,* rev. ed. Oxford: Newnes.

Woodbury, Gregory G. 1994–1995. "Net Cultural Assumptions." *Amateur Computerist* 6: 2–3.

Woodbury, John R., Stanley M. Besen, and Gary M. Fournier. 1983. "The Determinants of Network Television Program Prices: Implicit Contracts, Regulations, and Bargaining Power." *Bell Journal of Economics* 14: 351–365.

INDEX